W9-ACX-249

Landmarks in Western Science

From Prehistory to the Atomic Age

Title page of Vesalius's *De Fabrica*, 1543.
(The British Library, C.54.K.12)

Landmarks in Western Science

From Prehistory to the Atomic Age

&❧

Peter Whitfield

Routledge

New York

509
W588l

Published in the United States of America in 1999 by

Routledge
29 West 35th Street
New York, NY 10001

Published in Great Britain in 1999 by
The British Library
96 Euston Road
London NW1 2DB

Text copyright © 1999 by Peter Whitfield

Library of Congress Cataloging-in-Publication Data
Whitfield, Peter
  Landmarks in western science : from prehistory to the atomic age /
Peter Whitfield
   p. cm.
  Includes bibliographical references and index.
  ISBN 0-415-92533-9 (alk paper)
  1. Science–History. I. Title.
Q125.W59 1999
509–dc21   99-24976
                                            CIP

Designed and typeset by Peter & Alison Guy
Printed in Hong Kong

All rights reserved. No part of this book may be reprinted or reproduced or utilized in any form or by any electronic, mechanical, or other means, now known or hereafter invented, including photocopying and recording or in any information storage or retrieval system, without permission in writing from the publishers.

Routledge, Inc. respects international copyright laws. Any omissions or oversights in the acknowledgements section of this volume are purely unintentional.

PACKARD LIBRARY

JUL 0 1 2003

THE COLUMBUS COLLEGE
OF ART AND DESIGN

Contents

	Preface	7
Chapter One	The Origins of Recorded Science	9
Chapter Two	The Classical Achievement	28
Chapter Three	Science in Religious Cultures	
	Part 1. The Islamic Masters	50
	Part 2. Christian Pupils	67
Chapter Four	The Problem of the Renaissance	90
Chapter Five	Science Reborn	120
Chapter Six	Eighteenth-Century Interlude	160
Chapter Seven	The Machine Age	182
Chapter Eight	Twentieth-Century Science: the New Labyrinth	225
	Select bibliography	247
	Index	251

All life must pass, in Shakespeare's compelling phrase, 'through nature to eternity', and the mystery of that transition might be regarded as the subject-matter of science. The discoveries of modern physics have provided an almost uncanny vindication of Shakespeare's words, for we now know that through all nature's forms and processes, its substance is eternal, and that particles formed billions of years ago have passed from interstellar space to our planet, into our bodies, from where they will return again to their cosmic source. This understanding of nature - as a succession of forms assumed by eternal elements - emerged during the hundred years between 1840 and 1940, when theories of thermodynamics, atomic structure, and the equivalence of matter and energy all took shape, and it stands as arguably the greatest single achievement of human thought. This book is an attempt to describe the long and complex process of discovery that lies behind that achievement, and to show how man has tried for centuries to build bridges between nature and eternity.

Today we define science by its methodology, its critical examination of the material world, its language of measurement, experiment and deduction, and its attempt to formulate general laws which describe the way things behave. But this view of science is largely the creation of the last two centuries. For thousands of years, in a myriad of different cultures, science would have been defined not by its method but by its content, that it was engaged in constructing a set of beliefs about nature and its workings. For this activity, 'natural philosophy' was a better term than science, indeed it was still the term used by Newton in the late seventeenth century in the title of his great work of classical physics. Seen in this light, the history of science becomes an inclusive discipline that is central to man's entire intellectual history. The historian of science cannot proceed by searching the past for anticipations of modern scientific ideas; instead he adopts the model of science as answering questions about our world and our place in it, and asks how these questions were answered in the past. How have men explained the richness of the living world, the processes of disease and death, the movements of the heavens, the cycle of the seasons, and a thousand other questions? Problems such as these have defined natural philosophy throughout recorded history, and although the way they have been answered in the past would now be seen as religious, mythical or poetic, yet those answers are still a part of the history of science. This historical study leads inevitably to a point of fundamental interest: when did 'scientific thought' as we know it first emerge and become dominant over other forms of thought about nature and man? The difficulty in finding an exact answer to this question suggests that the history of science, like that of art, is not a simple progression from lower to higher, but a sequence of responses to the world, conditioned by historical circumstances, and having the central question of nature always at its heart.

A book as short as this about a topic so vast calls for some apologies. I am keenly aware that every subject and individual discussed here should be dealt with at far greater length, if justice were to be done to them; but this would have resulted in a very different and less accessible book. My only excuse for writing so briefly is that I had searched for years for a book like this - for an outline of scientific history - and could never find one. Medicine has been treated quite generally because it has such an extensive literature of its own; likewise the development of mathematics, and its interaction with physical science, is outside the scope of this book. If I have seemed to discuss philosophy and religion as often as strict science, it is because knowledge has never existed in a vacuum, and the scientists of the past have often sought to place their work within a philosophical framework. It was not a scientist but a theologian, Karl Barth, who wrote 'heaven is the creation inconceivable to man, earth the creation conceivable to him; he himself is the creature on the boundary between heaven and earth', but few great scientists would have disagreed with him. Our modern knowledge of nature's intricate forms and processes is now immense, but this knowledge is only half the story. More and more, the ancient question of purpose is being revived: *why* does nature function as it does, and not otherwise? Are the systems that exist inevitable, or could others have evolved? Is there a design at work, and if so what is its source? Has the human intellect some special part to play in recognising, or perhaps even modifying, this design? In some four thousand years of recorded history, science has travelled an immense distance, but to answer these fundamental questions, it may have to travel as far again.

I wish to thank John North for the part he has played in the preparation of my book, both through the example of his own writings, and through his criticisms of this text. Kathy Houghton of the British Library carried out much of the picture research. Finally thanks to my son, Eliot Whitfield, for introducing me to Verne's Professor Lidenbrock, the perfect symbol of the heroic element in the history of science.

P.W.

Chapter One

THE ORIGINS OF
RECORDED SCIENCE

&❦

'The universe is a device for making deities'
Henri Bergson

THE PREHISTORIC PERIOD

Scientific knowledge lies at the heart of human history, for it is man's ability to recognize nature's patterns, laws or regularities and to use them for his own purposes which marks him out from all other species. In this process technology preceded science, and it was the act of shaping and using tools – at least two million years ago – which defined the dawn of human history. Much later, the origin of 'civilization' was associated with agriculture and metallurgy, crucial skills in the manipulation of the human environment. Today we think of science as an intellectual edifice, a quest for the laws which govern the universe. For thousands of years however, man was content to use empirical knowledge and skill without seeking to articulate what we would recognize as a rational basis for them. Ancient cultures could display enormous technical inventiveness, but scientific thought is harder to recognize. The human mind has always sought for explanations and causes behind the regularities, or indeed the upheavals of nature – the pattern of the seasons, the birth and growth of plants or animals, the cycles of Sun and Moon, catastrophic storms, mysterious disease or sudden death. For the larger part of human history however, the drive to explain led to the realm of religion and myth rather than science, for nature was believed to be the arena where divine powers were at work. In this sense, man's earliest intellectual impulse foreshadowed what was to become the perennial quest of science: to discover what is is hidden, to uncover the controlling forces that lie behind what we see.

That nature's processes and mysteries exercised men's minds from the earliest times is clear from even the fragmentary record of prehistoric cultures: the burial of the dead and the attendant signs of belief in an afterlife; the carving or painting of human and animal figures; the embodiment of astronomical knowledge in stone structures. Although these things were products of pre-literate societies, so that we cannot grasp their exact significance, they spring undeniably from human thought about nature and its enigmas.

The precise dating of man as a tool-user is fraught with difficulty. No doubt for many thousands of years early hominids made use of material objects to extend their strength or reach, before taking the crucial step of making tools for themselves, and thousands more must have passed before they developed the skills to design tools for specific uses. This degree of specialization (for example the making of composite tools where stone blades were mounted in wooden hafts) naturally occured at different times across Africa and Eurasia, but with it invariably

Trepanned skulls of the
Neolithic period.
Whether these operations
were carried out for
cultic or medical purposes
is unknown, but the
healed wounds indicate
that the subjects lived
and recovered.
National Museum, Copenhagen

come definite signs of cultural enrichment, with evidence of social customs and beliefs, which in turn suggest the existence of language. Among these social customs was the care of the sick, with clear evidence of the healing of broken bones, and of individuals living many years after limb-amputations or skull fractures. One of the most enigmatic medical survivals from prehistory is the evidence of trepanning the human skull. Examples have been found from a number of neolithic sites, with indications that the wound has healed successfully. Whether it was carried out for medical reasons or was part of some rite, for example releasing the patient from demonic possession, can only be conjectured. The oldest known burials fall in the middle Palaeolithic period, between 50,000 and 100,000 years ago, and from the first the dead were interred with grave goods such as weapons, ornaments and, enigmatically, animal skeletons. It is difficult not to interpret these practices as evidence of belief in an afterlife, but the exact nature these beliefs is beyond recovery. It is surely no accident that an enrichment of thought occurred simultaneously with a diversification of man-made artefacts, for the experience of tool-making was supremely important to early man, suggesting as it did the power of creativity and causation – that things happen because they are designed and willed by a conscious being. In this insight may lie the genesis of the belief in personal deities who direct the forces of nature. Evidence of animal sacrifice – by fire, by water or by burial – dates also from the middle palaeolithic, and points irresistibly to a belief in these forces, which presumably were regarded as personal and subject to persuasion and propitiation.

From this period too comes a further swathe of evidence of man's feeling for nature in the first works of art. The cave-paintings of Lascaux and elsewhere are perhaps 20,000 years old, but are so finished and mature that we must believe them to be the expression of a tradition which had been developing for millenia. The content of this art is the living world, in particular the powerful horned animals – bison, stag and ibex – which were hunted for their meat, their bone and their skin. Human activities, trees or other natural objects, the Sun and Moon – none of these things appear in this oldest known art-form. This striking fact, together with the hidden, barely accessible cave-locations of the paintings, suggest that these images were connected with some magical or ritual practices, possibly a form of hunting magic, in which the capturing of the beast's image foreshadowed that of the beast itself. Similarly, the pictures of humans wearing animal masks suggest either a ritual dance in which a beast was killed by proxy, or a form of victory dance celebrating the dominance of humans over the other species. These images of human shape-changing are intensely interesting as the earliest known examples of the personification of natural forces, which would lead in time to anthropomorphic deities – gods of rivers, sky, fire, disease or death, which, in mythologies the world over, would combine the human and the animal form. The creation myths which evolved in almost all cultures testify to an instinctive understanding of cause and effect in nature, and to the impulse to personify those causes.

Shaman in an animal mask, from a painting in Les Trois Frères cave, France, at least 20,000 years old. This ritual shape-changing may have been connected with the idea of deities who personified the forces of nature.

The human beings of the middle and upper Palaeolithic who crafted composite tools, cared for their sick, buried their dead, painted life-like images in their caves, and sacrificed to higher beings, were genetically identical to ourselves. Their thought-processes, while they could not yet embrace science in our sense, were so close to our own that they were irresistibly drawn to the subject-matter of science – to the realm of nature, the problems of life and death, control and causation, and the forces that governed the harsh, unforgiving world which they experienced.

At the end of the Palaeolithic era, around 10,000 years ago, there occurred one of the major cultural diversifications in world history – the development of agriculture in South-Western Asia (and slightly later in Meso-America and South-East Asia). This event permitted mankind to abandon the semi-nomadic hunter-gatherer existence that had been his for perhaps one hundred thousand years, and to form large settled communities with controlled, renewable food-supplies. This in turn led to incalculable social consequences, with division of labour, political structures, new technologies, and surplus wealth – the things we have come to regard as the marks of civilization. Among these new technologies were pottery, textiles, metallurgy and a new mechanics based on the wheel. There were vital advances in food production such as the technique of breadmaking, and the discovery of the properties of salt; the horse was tamed, wheeled carts were constructed, and grazing animals were domesticated. We have almost no hard evidence about how and when these techniques originated or how they spread; we must assume that they were all purely empirical discoveries, not systematic or scientific in the strict sense, yet they must have arisen from a keen observation of nature and a determination to master the environment. These discoveries meant fundamental changes in the conduct of human affairs, distancing man further and further from the animal struggle for mere survival, stretching his intellect into new realms of thought about his world and his place in it. In the course of several thousand years this revolution extended westwards across Europe until by around 4,000 BC, agriculture had reached, if not become dominant in, France, Britain, southern Scandinavia and Spain. One of the cardinal features of civilization – literacy – was still missing, but it is within the pre-literate

Bronze-age funerary mask
and hands from Graz,
Austria. They were affixed
to a coffin or urn, and
attempted to portray
the individuality of
the deceased.
Landesmuseum, Graz

culture of north-west Europe that one of the most intriguing forms of true science first emerged – the megalithic astronomy of Carnac, Callanish and Stonehenge.

Why are these megalithic sites so important in the history of civilisation? The answer lies in their scale: as feats of engineering created by the unaided strength of human hands, they represent a quite colossal investment of time, effort and ingenuity, amounting in the larger sites to millions of man-hours. Only the most powerful of motives can have inspired such projects, therefore if we can penetrate those motives these sites should offer an invaluable insight into the mind and culture of the men who built them. These sites were long regarded by antiquaries as cultic and ceremonial (or possibly defensive) and some astronomical or calendrical element was often imagined, for example if the number of stones happened to be twelve or thirty, or if it were seven then worship of the seven planets was invoked. But the idea that these structures embodied deliberate and precise alignments on Sun, Moon or stars was worked out in detail by the astronomer Sir Norman Lockyer who, at the end of the nineteenth century, founded an approach to these monuments which has generated innumerable theories, many of them conflicting, as to their origin and purpose. What was possible, according to the geometric analysis of the modern astronomer, and what actually happened on these sites four or five thousand years ago, may be two very different things. It may now be possible to demonstrate the alignment of certain stones on a certain celestial event at a determined time; but from these pre-literate societies there is of course no independent evidence that this event was in the builder's mind when he set up the stones. On the other hand, if a weight of evidence accumulates, as it has done, that astronomical alignments are traceable in many different but related sites, then we must take seriously the idea that astronomy was an important

The mysterious Newgrange
burial chamber, Ireland.
At the midwinter solstice,
the sunlight enters a single
aperture and illuminates
the tombs of the dead.
Dúchas, The Heritage Service,
Dublin

component in neolithic culture, although its ultimate purpose may still remain hidden.

These sites offer us certain elements in an immensely important jig-saw puzzle – that of the mind of ancient man – but how they fit together we do not know. At Newgrange in Ireland, a row of tombs was built in such a way that, at the midwinter solstice, the light of the rising sun enters the chamber corridor and falls upon a stone at the far end of the tombs. This is all the more striking because the main entrance to the chamber was deliberately blocked by a huge stone, and the light continued to enter through a narrow box above the entrance, long after living people could no longer do so. What was the association of ideas here? What was passing into (or perhaps out of) the tombs in these few moments of each year? In the past such a structure was likely to have been interpreted as in some sense calendrical, but a moment's thought will show that it was far too elaborate to have been designed for that purpose. What is true however is that such structures *embodied* calendrical events which had previously been studied and charted:

the alignments were of symbolic or religious significance, connecting beliefs about death with Sun, Moon or stars. In general terms such ideas still survive, having been incorporated into Christian architecture, where east is the sacred direction, and churches are built to face the rising Sun, while the dead also await the resurrection in their eastward-facing tombs. There are scores of other sites where some calendrical element has determined the structure, and this makes it all the more puzzling that no trace of any written calendar survives from this culture, inscribed on wood or stone. Could structures such as Newgrange have been built in a purely oral culture, their planning directed by the stored memory of an elite? In North-East Scotland there occurs a distinctive complex of stone monuments which create artificial horizons marking the highest and lowest declinations of the Moon which occur in an eighteen-year cycle. The observations that lie behind these monuments must have occupied decades of Moon-watching – and also memorizing?

Perhaps still more intriguing is the growing volume of evidence that before Sun and Moon came to dominate neolithic astronomy, certain monuments, including burial-mounds, were aligned upon bright stars. The burial mound at West Kennet (some fifteen miles north of Stonehenge) dating from the fourth millenium BC, contains five tomb-chambers in which the remains of forty-six individuals have been found. Each of the chambers, when the stones blocking their entrances are removed, receives the light from the rising of a bright star such as Betelgeuse, Spica or the Pleiades. As with Newgrange, the association of the dead with the light of celestial bodies is clearly implied, but the beliefs involved cannot be spelled out. There are many other examples where neolithic geometry appears to embody stellar astronomy, and where the sight-lines are correct for the celestial chart of that epoch, confirmed by carbon dating for example. It should be said however that many archaeologists do not accept that these stellar alignments are intentional or even authentic, and that if one takes twenty or more bright stars scattered across the sky, apparent alignments to them can be found which are actually random and without significance.

Such disagreements still plague the study of the most famous of all these structures, Stonehenge. That it was connected with some form of solar cult seems beyond doubt, but the long-cherished idea that the axis of the entire site points north-east to rising sun at the summer solstice is now known to be inaccurate. Sight-lines to the south-west, to the setting sun at the midwinter solstice are now considered more likely, but the motive and the actual ritual that was enacted there are still obscure. Perhaps the central point of interest in these cultic-astronomical sites is that they imply a belief that nature's disparate parts are related, that man in his earthly sphere was not isolated, but was linked to the realm of the stars. Perhaps that realm was believed to contain unseen powers which could be understood, and to some extent controlled in the interests of mankind. To achieve this, the patient observation of nature's patterns was essential so that future events could be accurately predicted. The whole activity was rule-directed and scientific in a very real sense, although the motive may appear to us to be religious: there may have been no distinction between the two realms of thought.

SCIENCE IN ANCIENT EGYPT

When we turn to the civilizations of the ancient near east, some of the difficulties in understanding the thought and the science of these early cultures are lightened – but certainly not solved – with the invention of writing, one of the great turning points in human history. Between 3500 and 3000 BC two distinct forms of writing emerged almost simultaneously, the cuneiform script of Mesopotamia and the hieroglyphics of Egypt. The survival of thousands of texts in these languages, puts our understanding of these cultures onto an entirely different plane from that of the cultures of neolithic Europe, which were exactly contemporaneous. For this reason the achievements of ancient Egypt and, more especially, of the Mesopotamian kingdoms, form the earliest chapters in the history of recorded science.

The familiar symbols of ancient Egypt – the pyramids, the hieroglyphics, the cult of the dead – are entirely appropriate, for in their different ways they all point to the enormous technical achievements which have defied the passage of thousands of years; but they point also to the limitations and the conservatism of Egyptian intellectual life. Hieroglyphic writing may have begun as a direct system of picture representations – pictures of a bird, an eye or pyramid meant precisely those things. But if this initial stage occurred, no evidence of it has survived and it would swiftly have revealed its inadequacy in dealing with conceptual words, even simple ones such as 'small', 'cold', 'pain', 'enemy', and so on. From its earliest appearance, the hieroglyphic system detached the phonetic values of the objects pictured and used them to build up a phonetic language. This was made simpler by the fact that only the consonants were used, the vowel elements being supplied by the reader. For example the consonant series m + r might be shown as a chisel or a pyramid, but the same consonants also formed the words 'to love' and 'to be ill', thus the same symbol might have two, three or four meanings, and a system of reading aids or determinatives was evolved to make the meaning clear in each context. In the early dynastic period the number of hieroglyphs totalled approximately 700. They were gradually increased by the addition of new signs down to the Roman era, but always in accordance with the principle of phonetic detachment; the switch to the more flexible and precise alphabetic system was never made. Because of the graphic, artistic nature of hieroglyphics, a simplified form was found necessary which could be excecuted more swiftly in a cursive form; the resulting hieratic script follows the hieroglyphic principle exactly. The survival of this system of writing for three thousand years into the era of Greek, Latin and other alphabetic languages, testifies to the conservatism of Egyptian culture.

The advent of writing was a momentous event in the history of mathematics, for the working processes that could be retained in the memory were obviously limited, and the Egyptians made good use of the new medium of calculation to create an effective, if slightly cumbersome form of mathematics. The numerical system was a decimal one in the sense that ten and its multiples were treated as significant divisions, and additions and subtractions were arranged in columns of hundreds, tens and units, with results being carried from one column to another. On the other hand each multiple of ten had its own written symbol, and there was no positional element in the number's value, as there is with the modern decimal system. Multiplication was

The Origins of Recorded Science

Egyptian numerals. Mathematics could not develop without a flexible notation; the Egyptian system used multiples of ten, but it was not a decimal system in the modern sense, since there was no positional element.
from *The Dictionary of Scientific Biography*

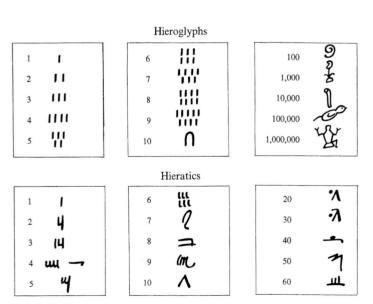

Hieroglyphs

1	I	6	III / III	100	
2	I I	7	IIII / III	1,000	
3	III	8	IIII / IIII	10,000	
4	IIII	9	IIIII / IIII	100,000	
5	III / II	10	∩	1,000,000	

Hieratics

1	I	6		20	
2		7		30	
3		8		40	
4		9		50	
5		10	∧	60	

performed by a process of repeated addition, thus 9 times 26 becomes: 26, 52, 104, 208, 234. Division was effected by reversing this process, thus 234 divided by 26 becomes the problem 'By what must I multiply 26 to reach 234?' The Egyptians were well versed in the use of fractions, but with the exception of $^2/_3$, wrote them always as unit fractions, thus $^4/_7$ would be written as $^1/_2 + ^1/_{14}$, while $^2/_{29}$ becomes $^1/_{24} + ^1/_{58} + ^1/_{174} + ^1/_{232}$. Such fractions could clearly be expressed in a number of different ways, and there seems no satisfactory explanation why this complex system persisted and why the scribes failed to employ numerators other than one.

This undoubted facility with numbers was used to good practical effect in solving problems of area, unit parts, weight and volume. A mathematical papyrus of *c.*1850 BC gives many problems and the formulae for solving them, such as calculating the volume of a frustrum – a truncated pyramid:

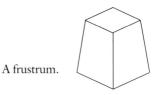

A frustrum.

'If it is said to thee, a truncated pyramid of 6 cubits in heights,
Of 4 cubits of the base by 2 cubits of the top,
Reckon thou with this 4, its square is 16.
Multiply this 4 by 2, result 8.
Reckon thou with this 2, its square is 4.
Add together this 16, with this 8, and with this 4, result 28.
Calculate thou $^1/_3$ of 6, result 2.
Calculate thou with 28 twice, result 56.
Lo ! It is 56. You have correctly found it.'

We could express this by writing *a* for the base of 4 cubits, *b* for the top of 2 cubits, and *h* for the height of 6 cubits; then we find that the scribe has correctly calculated the volume of the frustrum by $V = ^1/_3 h\ (a^2 + ab + b^2)$. How the Egyptians discovered this quite sophisticated formula however, we have no means of knowing. There is evidence among these problems of a type of algebraic reasoning to discover unknown quantities, for example if one fifth of a quantity is added to the quantity and the result is 20, what is the quantity? We would express this as $^x/_5 + x = 20$, but this species of algebraic notation was unknown to the Egyptians. Another papyrus of a similar date shows that the Egyptians were drawn to the problem of the circle, and gives the procedure for calculating the area thus: 1/9 of the diameter is discarded and and the result is squared. For example if the diameter is 9, the area is 64. This recognizes that the area of a circle is proportional to the square of the diameter, and that the constant of that proportion is 64/81; we would call this $\pi/4$, thus giving a quite accurate value for π of 3.160. Is it possible to relate such mathematical activity to colossal practical tasks like building the pyramids?

The problem of the pyramids exists on several levels: were they mathematically planned in detail and in advance? How were the enormous weights handled? And what was the motive for such a vast expenditure of human energy? The Great Pyramid at Gizeh dating from 2800 BC has sides averaging 755 feet 9 inches long at the base; two of its sides vary from that figure merely by inches. Its facing consists of more than two million limestone blocks averaging $1^1/_2$ tons in weight, which are carried upwards at a constant angle of 51° 51' for over two hundred courses, reaching a height of almost 500 feet. This structure would tax the skill of the modern engineer armed with every conceivable power tool. The Egyptians seem not to have been familiar with screws or pulleys; the horse and the wheel were unknown to them before *c.*2000 BC; therefore we have no other clues to their means of construction than the few surviving images of rank upon rank of tiny figures hauling massive stones into place by hand. That ramps or roadways were raised up to each level seems indisputable, although no clear evidence for them has been found. Beyond the engineering problem however is the enigma of motive and purpose: why did a dead body, even that of a king, require such a stupendous resting-place?

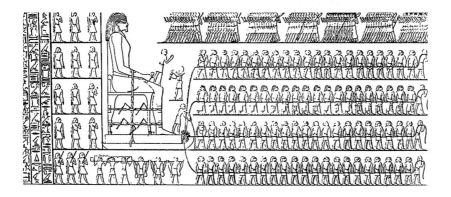

Moving an Egyptian statue, from a bas-relief. This does not tell us how Egyptian monuments were constructed, but it does show that unlimited human power was used, rather than mechanical aids.
after Lepsius *Denkmaeler*, 1848

With the help of texts and inscriptions we can attempt an answer, but it will be tentative for these texts present us with certain beliefs and practices, but not with the kind of explanations which would tie them together into a philosophical whole. The Egyptian king, while a mortal human, was undoubtedly accorded divine status too. Of his many titles the most important was the Horus-title: he was the incarnation of the falcon-headed god whose two eyes were the Sun and Moon; later he was also called son of Re, the Sun-god. Re was also the father of Ma'at, goddess of order and justice, and thus the king was among the pantheon of the gods, a part of the unchanging cosmic order. His role was to mediate between gods and men and to uphold the principle of Ma'at, the order which was believed to underlie both the cosmic and social structure. Clearly then, although subject to natural death, he could not cease to exist, but must live on in another realm. Thus the dead, at least the royal dead, were as much a part of the cosmic order as the living. Although the Egyptians believed in spiritual components of human personality which survived death, they also believed that the body must remain intact so that the dead could continue to live in their world. It is this rather paradoxical doctrine which alone can explain the massive architecture of the dead which the pyramids represent. We are justified in asking at what social level these beliefs operated: were they the preserve of a royal and priestly elite? They certainly cannot have been shared by the tomb-robbers who penetrated even the formidable defences of the pyramids, forcing the court to seek more secretive tombs in the Valley of Kings. There seems little doubt however that the breathtaking engineering of the pyramids was directly inspired by beliefs about cosmic order, and this fact must shed light on the contemporary monument-building cultures which left no written records, such as neolithic Europe. The Gizeh pyramids are orientated precisely on the cardinal compass points while certain galleries within the pyramids appear to be aligned on the Pole Star of that epoch, Thuban. None of these matters is explained or even referred to in any surviving text, and their religious significance is therefore still highly uncertain.

There is a second sense in which the cult of the dead is a meeting-place of religion and technology, namely the practice of embalming. It is possible that the Egyptian custom of wrapping the dead in cloth and burying them in arid soil was influential in fostering the doctrine of the afterlife, for the form and features would be preserved in this way for a considerable time, encouraging the belief that life still resided there. As in Christianity, a belief in an afterlife of the spirit co-existed with an expectation of physical resurrection: for this to happen it was considered essential that the elements of the whole person – soul, name, shadow, heart and body – were reunited. It was the need to preserve the body's form and beauty, in order to lure the soul back to it, that lay behind the techniques of embalming, which were sophisticated and costly. The internal organs including

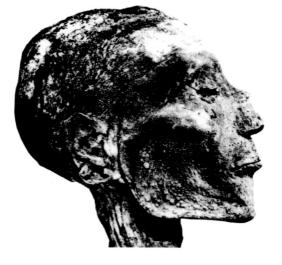

Mummified head of the Egyptian King, Rameses V, who died in 1160 BC. The skin shows lesions which are almost certainly those of smallpox.
Cairo Museum

the brain were removed by surgery and preserved in separate jars with wine and herbs. The body cavities were filled with preservative herbs and spices such as myrrh, and the whole body was placed for some months in saltpetre, after which it was swathed in gummed bandages and entombed. That such macabre mortuary practices should have a theological motive has always baffled the western mind, since Herodotus gave the first detailed account of the technique.

It might be expected that these customs would promote Egyptian knowledge of anatomy and medicine, and to a certain extent this is true. Egyptians viewed the body as a system of channels, perhaps by analogy to the channels which carried life-giving water through their land. The heart was considered to be the centre of this system and its pulsating action throughout the body was recognized. However the channels were believed to carry not only blood but other fluids such as urine, tears and sperm. Although the individual organs were familiar, opinions about their function were confused, for example air was thought to enter the body through the mouth and nose, but also through the ear, and to travel directly to the heart and thence throughout the body. The importance of the brain was recognized, and one medical papyrus describes the effects on motor functions of brain injury, and the left-right correlation was noticed. As with most societies, there was a body of knowledge concerning natural drugs and herbs, and their effects on specific symptoms, such as fever, rash or sickness, but naturally there was no classification of diseases: the symptom was the disease, and the physician's skill was directed against the symptoms. Surgery is barely mentioned in the surviving medical papyri, but cauterization of cysts and tumours was common. The practice of embalming has incidentally provided modern medicine with a rich field for historical diagnosis. A high proportion of mummies show evidence of quite serious disease of the skeleton, as well as parasitic infestation and smallpox. Herodotus reported that 'the whole country is full of doctors', and it is certainly true that the Egyptians have left considerably more information about their medicine than about their physical sciences.

The applied science of ancient Egypt reached considerable heights in one other field, that of astronomy used in the service of calendar-reckoning. The highly-organized, hieratic society of ancient Egypt demanded an effective calendar for its civil and religious life, and through a tradition of meticulous observation of Sun, Moon and stars, created a calendar whose structure has been praised as 'the only intelligent calendar which ever existed in human history'. However there are no astronomical texts from ancient Egypt which discuss matters such as the motions of the heavenly bodies, no evidence that the structure of the cosmos was seen as a field of intellectual enquiry. The annual event of the greatest importance for Egypt's life was the flooding of the Nile each July, which irrigated the surrounding land and made it fertile. It was noticed that this event was immediately preceded by the first pre-dawn rising of Sirius, the brightest star in the sky, which the Egyptians called Sothis. This date became the cornerstone upon which a calendar year was constructed consisting of three seasons: Flood, Emergence and Low Water or Harvest. These seasons in turn were divided into three lunar months, each containing a religious festival after which it was named. Of course this stellar event did not coincide with the lunar divisions, and various intercalary periods became necessary, used on a flexible basis. This was acceptable for the placing of festivals, but became inconvenient for the civil administrators, and a second calendar was devised, fixed to the natural year of 365 days, but divided into twelve months of thirty days, with five intercalary days regularly added at the end of each year. This did not mean that the stellar-lunar calendar was abandoned; the two systems ran simultaneously, maintained by a professional class of astronomers. Of course the fact that the natural year is actually close to 365.25 days and not 365, meant that the civil year would gradually move forward through the natural year, and the Egyptians well aware of this, indeed they calculated that it would require 1,460 years before the two years were again in precise agreement – the so-called Sothic cycle. It says much for their skill in calendar-making that they could handle these conflicting systems so expertly.

The second aspect of time-division which the Egyptians inaugurated and which has become permanent is the twenty-four hour day. The division of the year into 365 days has a natural astronomical basis, but that of the day into twenty-four hours is quite arbitrary. Although the choice of twenty-four is not directly explained, there is evidence that it was based on their belief that the Sun-god, Re, spent the hours of the night voyaging through the underworld in a boat with his divine companions. The underworld was said to be divided into twelve regions, and the god must pass through certain dangers and progress through the intervening gateways, passing one hour in each; the word for hour was written as a star. Once again using the principle of star-rising, the astronomer priest selected twelve bright stars which rose in succession through the night, and, calling the interval between each an hour, charted the progress of the Sun-god through the underworld. In practice it was not a single star which was used but a group, a constellation in fact (although they were not our constellations), and suitable star groups visible at different times of the year were selected, totalling thirty-six. These thirty-six star-groups were known as decans, since each one rose heliacally (that is the first pre-dawn rising) for ten days. This whole system is made explicit in the painted star clocks which were prevalent from around 2200 BC, in which twelve star-groups are charted for each ten-day period of the year. The division of the day into twelve hours was made by analogy with the night, and the the daylight hours were measured by shadow-clocks. These hours would of course have

Egyptian cosmology: the sky-goddess Nut, bends over the earth-god Geb. A creation mythology involving such deities co-existed with the meticulous study of the movements of celestial bodies, upon which the calendar was based.
Redrawn from a papyrus of the 10th century BC

been of unequal duration, because of the seasonal variation of night and day, and there is evidence from around 1500 BC of the introduction of water-clocks, in which water leaked slowly from a vessel, to give equal hours.

In addition to the thirty-six decanal constellations, the Egyptians also designated a group of a dozen northern constellations which were always visible in the circumpolar sky. Curiously, none of these constellations corresponds to our own, with the sole exception of the Big Dipper section of Ursa Major, which is shown as a bull. The five naked-eye planets were observed by the Egyptians and were identified with certain gods, but no special interest seems to have been displayed in their irregular paths, and no special religious significance seems to have been attached to them. Rather strangely to our eyes, all three outer planets – Mars, Jupiter and Saturn – were considered to be aspects of Horus.

Political events in the sixth century BC inaugurated an era of extensive change in Egyptian astronomy. The conquest of Egypt by Persia brought Egypt into contact with the scientific traditions of Mesopotamia, and we know that individual scholars visited the ancient cities of the Persian Empire and brought back new astronomical concepts. Among the most important of these were the Zodiac, a heightened interest in the planetary deities, and judicial astrology, which provided a new context for astronomical activity. From the Graeco-Roman period, celestial charts survive in which the twelve Zodiacal constellations are integrated with the thirty-six decans, and astrological texts begin to be written in which the fate of kings and nations is foretold from conjunctions of the Sun, Moon and planets and from events such as eclipses, which had not previously concerned the Egyptians. This aspect of astronomy was eagerly developed, and it was at this time that Greek travellers such as Herodotus began to visit Egypt and to report this entire school of omens, divination and astrological conjunctions as a creation of the Egyptians, ignorant of its true origins in the kingdoms of Mesopotamia.

The Egyptian sciences that are best recorded, mathematics, medicine and calendrical astronomy, are practical in application and conservative in character – two thousand years saw no revolutions and few changes of note. The spirit of this tradition is captured in the words which Harkhebi, an astronomer of the third century BC, had inscribed on his memorial:

'... Wise in the sacred writings, who observes everything observable in heaven and earth, clear-eyed in observing the stars, among which there is no erring ... who observes the culmination of every star in the sky, who knows the heliacal risings of each ... who foretells the heliacal rising of Sothis at the beginning of the year ... who divides the hours of the day and night without error ... knowledgeable in everything which is seen in the sky, for which he has waited, skilled, with respect to their conjunctions and their regular movements ...'

Harkhebi's description of his role is limited to observing and recording; there seems to be little evidence of an over-arching conception of nature, its laws or its mysteries. On the other hand we cannot be sure what the 'sacred writings' may be to which he refers, and the Egyptian cosmic myths and their cult of the dead suggest deeper dimensions of thought, which have left few explicit traces in the written legacy of their civilization.

SCIENCE IN MESOPOTAMIA

Just 500 miles of desert separated Egypt from Mesopotamia, that other centre of civilization, where writing, science and technology created a new phase in the history of mankind. It is from around 3200 BC that the first evidence survives of the cuneiform script which was to be developed by the kingdoms of Assyria, Babylon and Persia during the next three thousand years. The region of Mesopotamia knew many masters, and its cultural history passed through many phases, some far richer in written in records than others. Between 3000 BC and 500 BC, we find records of Sumerian, Akkadian, Babylonian, Assyrian and Persian science. These peoples all employed the cuneiform script, which is thought to have originated as a series of pictures which were simplified into wedge-shaped strokes adapted to the writing medium, which was wet clay. As with hieroglyphics, the pictograph became shifted from its original subject to its phonetic component, so that 'hand' might become H, and so on, although the cuneiform symbols more often became syllables than letter-equivalents.

There are some significant differences in the scientific thought of these various periods, but the central Babylonian-Assyrian tradition from *c.*1800 BC shows a continuity and a self-sufficiency of which its practitioners were proud. This is an oration spoken by the Assyrian king Ashurbanipal about the year 650 BC:

'I learned the hidden wisdom, the whole art of the scribes
   and astrologers,
I can intepret omens of heaven and earth,
I participate in the councils of experts,
I can discuss treatises of divining with the skilled seers,
I can find difficult reciprocals and products which are not
   easy to calculate,
I can read elaborate texts in which the Sumerian and Akkadian
   are obscure,
I can decipher stone inscriptions which date from before the flood.'

This text is highly revealing of both the subject-matter and the context of Babylonian science: its reiterated concern is with acquired wisdom in mathematics, astronomy and divination. The latter was central to the thought of the Babylonians,

Ashurbanipal, the Assyrian
king who boasted of his
skill in mathematics,
astronomy and divination.
He is returning from the
hunt, and consecrates to
his gods the lions he
has killed.
British Museum WA 124886-7

for they believed fervently that the cosmic order interacted with the human, and
that the gods wrote intimations of the future into the natural world for man to
read. The experts referred to here included men who foretold the future by studying the pattern of bird-flight in the sky, by inspecting the organs of sacrificed
animals, by watching the behaviour of oil in water. These were the experts with
whom Assurbanipal claimed equality, and who used their skills to foretell the fate
of military campaigns, or of the progress of sickness. Babylonian science was in
the hands of these seers, astronomers and physicians who constituted an intellectual elite centred on the royal court. Their wisdom was contained in long,
detailed formula-texts in which all possible signs and permutations were
described and interpreted, thus:

> 'If the oil I have poured on the water has descended then risen, and has
> surrounded the water: for the military campaign the arrival of calamity;
> for the sick man the hand of god, the hand is heavy.'

In this text thirty or more patterns of oil in water and their meanings are given,
while in medical texts hundreds of localized symptoms are given for every part of
the body from skull to foot ('If the skin of the left arm/right arm/left hand/right
hand' etc.is 'raised/shrunk/whitened/rough' etc.), the diagnosis is not that the
patient has a named disease, but that the sickness will vanish or worsen. These
formulae seem to be the opposite of our search for general laws that may be
applied in different cases, but show a desire to tabulate every possible outward
symptom, thus incidentally displaying the practitioner's mastery of his subject;
there could be at this stage no systematization by disease. This exhaustive listing,
this culture of formulae, is found also in the celebrated law-code of Hammurabi,
where a multiplicity of offences are spelled out and their penalties are matched to
the age, social class, relationship, or sex of those involved. The considerable number of surviving texts enable us to build up quite a full picture of Babylonian
medicine. The overriding belief was that disease was caused by divine disfavour,
or demonic possession. This did not mean that the sick were not treated, or were
thought to be beyond human help, but it did mean that no functional theory of
disease emerged, and treatment was directed entirely to the symptoms. Several
different professional groups were involved with medicine: the diviner, the exor-
cist and the physician, and the practice of hepatoscopy – the examination of the
liver of sacrificied animals – was the most important single means of diagnosis.
Surgery was common as we know from the code of Hammurabi, which pre-
scribes the fees for a physician, but also the penalties for failure:

> 'If a doctor has treated a man with a metal knife for a severe wound and
> has caused the man to die, or has opened a man's tumour with a metal
> knife and destroyed the man's eye, his hands shall be cut off.'

The sick were kept as isolated as possible, not so that the healthy should avoid infection but so that they might escape the influence of the responsible demons.

What lay behind the Babylonian culture of divination? Why should the flight of birds foretell the outcome of a battle, or the organs of a dead sheep presage the death of a king? As in the case of Egypt there is a scarcity of philosophical texts explaining the basis for these beliefs, but it undoubtedly proceeded from a belief in the order and unity of nature, which was determined by the gods. The place of human events within the cosmic scheme became comprehensible if one read the guiding signs which the gods had written into certain natural processes. As one text states, 'Sky and earth both produce portents; though appearing separately, they are not separate for sky and earth are related'. Divination was clearly not a science in our sense, but it was a philosophy of nature, and in Babylonian culture it co-existed with a high order of achievement in the precise sciences of mathematics and astronomy. The stars were called 'the gods of the night', but their movements were the subject of intense study and record. Before the Persian period however, Babylonian religion was not wholly an astral religion, although it was cosmic: alongside the deities of Sun and Moon were gods of water, storm, fire and so on, while the creation epic *Enuma Elish* tells how the god Marduk overcame the monster of primal chaos, Tiamat, and cut her body in two to create the earth and heaven, causing the rivers Tigris and Euphrates to flow from her eyes.

The Babylonians inherited the Sumerian number system, a sophisticated system using place-value notation as we do with 10, but they took 60 as the base.

Bronze figure *c.*1000 BC of Pazuzu, the Babylonian demon thought to bring sickness.
The Louvre, Paris

Above, left. Babylonian clay model of a sheep's liver, *c.*1800 BC. It is inscribed with texts interpreting as omens blemishes which might appear on different parts of the organ, and was used as a reference when divining the future from the organ of a sacrificed animal.
British Museum WA 92668

A wedge represented 10 and a simple stroke 1, thus > > >∏ denoted 32, but if the symbols are appropriately spaced, the values change enormously, for example:

60²	60	1	¹⁄₆₀	modern equivalent
		> > >II		32
	III	> >		180 + 20 = 200
>	II	>		36,000 + 120 + 10 = 36,130
			> >	²⁰⁄₆₀ = ¹⁄₃

This looks difficult to grasp at the outset, perhaps because only two symbols are used, but when mastered it was a powerful system. The use of **60** as a base has survived in our measurement of time and of the angles of circles, an application transmitted through Egypt and Greece. Not surprisingly the Babylonians compiled tables to aid their calculations – tables for multiplication, reciprocals, squares and square roots. They developed more sophisticated methods than the Egyptians in solving equations, methods which we could term algebraic in that they manipulated unknowns, but they lacked an algebraic notation, using instead step-by-step procedures. These problems were often given geometric form:

> ' I have subtracted the side of a rectangle from the area, and the result is 870, what is the side?
> – Take one half of 1, result ¹⁄₂.
> – Multiply ¹⁄₂ × ¹⁄₂, result 1/4.
> – Add ¹⁄₄ to 870, result 870 1/4.
> – The square root of this is 29¹⁄₂.
> – Add ¹⁄₂ to 29¹⁄₂, result 30
> – 30 is the required side of the square'

We could express this problem as $x^2 - ax = b$, where $a = 1$, the solution to which is

$$x = \sqrt{(\tfrac{1}{2}a)^2 + b} + \tfrac{1}{2}a$$

Babylonian geometry dealt with areas of triangles and volumes of solids. The area of a circle was calculated by $3r^2$ and the circumference by $6r$; the π constant seems to have been unknown to them, and both these simplified formulae therefore give results approximately 5% in error. Less celebrated for their building than the Egyptians (partly because of the scarcity of good building stone), the Mesopotamians became practised hydraulic engineers, putting their mathematics to practical use in irrigation, damming and flood-relief. The structure of the characteristic ziggurat – the stepped pyramidal tower – was less colossal than the Egyptian pyramids, but still sufficiently imposing to inspire the biblical story of the Tower of Babel which attempted to reach heaven. The ziggurats were built as temples, but were almost certainly used as astronomical observatories too.

Astronomy was the field in which Babylonian mathematics and their rage for divination met to form a rich and immensely influential strand of natural philosophy. From the Old Babylonian period (*c.*1800 BC) onwards, a number of astronomical omen texts survive of a simple kind, many of them relating to the planet Venus, predicting good or bad harvests, plague, adversity for the king and so on. These general predictions constitute an early form of judicial astrology, as distinct from personal horoscopes which did not exist at this time. Venus was identified with the goddess Ishtar, and it should be emphasized that at this stage the effects predicted were believed to stem from the will of the goddess, her power being greatest when the planet was in the ascendant. The planet's position was then an index of divine power. The idea that the planets could affect human affairs *as planets* rather than as deities – the view in the west in middle ages and the Renaissance – is a later and, in a sense, a less rational one. By *c.*900 BC these omens grew to number several thousand, and concerned the Moon and the other

Mul Apin - a cuneiform star catalogue from Babylon, dated around 500 BC. It lists constellations and prominent stars, with the dates of their earliest pre-dawn risings. British Museum WA 86378

planets, following the 'exhaustive-list' pattern described above. A notable absence at this stage is the division of the sky by the figures of the Zodiac. Around 700 BC an important astronomical compendium was written entitled *Mul Apin* (the title meaning simply 'The Stars of Apin' – a constellation) in which the dates of rising and setting of some thirty-six constellations are described, together with their bright stars. They do not correspond entirely to our constellations, but many of them are related. The *Mul Apin* sky catalogue was evidently founded on years of systematic observation, and marks a new maturity in Babylonian astronomy. The structure of the Zodiac is not made explicit, but the authors knew that the Sun followed a path which carried it for three months to a high altitude, three months at a low altitude, and two periods of transition. They knew that between these two extremes lay also the paths of the Moon and the planets, and they identified a number of constellations lying on this path such as Virgo, Aries, Leo and Taurus. Thus the makings of the Zodiac and a celestial coordinate system existed in embryo in *Mul Apin*.

It is in texts from the later, Persian phase of Mesopotamian history that this vital development is found, when the sky is divided into twelve zones, each of thirty azimuthal degrees, and with all the planetary and lunar activity concentrated in a band twenty-three degrees either side of the celestial equator. Why was it so important? Because it created for the first time a rational map of the sky, within which celestial positions could be charted in the present, past or future. The motives for this may have been in part intellectual, but it coincides, whether as cause or effect is uncertain, with the beginnings of horoscopic astrology, in which the destiny of the individual is claimed to be discernible from the aspect of the heavens at his birth or at other times of crucial decision. To read this balance of powers among the heavenly bodies, some accurate positioning method was clearly essential, and the zoning of the sky into azimuth and altitude provided that framework. This form of divination may have been a special case of the concept of cosmic unity which underlay the older and cruder methods. However it may be more specifically related to the Persian doctrine, common to Mithraism and Zoroastrianism, that the soul's original home was among the stars, and that although now joined to the human body, the human personality is still governed by the stars. Whatever its precise sources, the rage for astrology spread from Babylon to Egypt, Greece and Rome, and remained a major intellectual force for some two thousand years.

It should be emphasized that this charting of the sky did not result in celestial maps as we know them, but took place on an entirely mathematical level. The Mesopotamian tradition never seems to have produced any mature model of the cosmos, corresponding to the celestial sphere of the Greeks. The plotting of celestial positions by spherical geometry was unknown to them, and instead they employed an intricate method of following the velocity of celestial body against time, yielding a path which we should plot as a graph, a linear function of time. Using their mastery of this method, they traced the positions of Sun, Moon and planets with enormous accuracy and entire cuneiform tablets were devoted to each body, carried many years into past and future. That this was more of a mathematical than an observational exercise, is demonstrated by the lunar tablets, in which full moons are charted even when they are invisible because the Moon is below the horizon.

In selective fields, Mesopotamian and Egyptian science reached a high level. But certain other realms of thought are strikingly absent: despite of centuries of meticulous sky-watching, there is no theory of the shape or mechanism of the cosmos; an intimate knowledge of the human body produced no theory of the functions of its organs, or even the concept of functional disease; the ubiquity of physical working against mass and weight did not evoke any language of physics. Their philosophy of nature led to them explain the world they saw by creating a parallel world that was unseen, a world whose laws integrated the realms of gods and men, life and death; in other words their science and their religion formed one intellectual structure. As D'Holbach, the eighteenth-century atheist, would express it: 'Failing to understand Nature, man created the gods'. Perhaps one revealing index of the limitations of their science is to be found in their maps of the world for both Egyptians and Babylonians have left so-called world maps which are in fact maps of their own regions bounded by oceans; *the* world was equivalent to *our* world, and neither logic nor experience could suggest to them any ways of enlarging this conception. Their legacy is not clear-cut and direct, as that of Greece is, but they invented the first written languages and the first systems of mathematics, and for this reason their intellectual history is of the highest importance.

Chapter Two

THE CLASSICAL ACHIEVEMENT

&❦

'Nothing exists except atoms and empty space;
everything else is opinion'

Democritus

THE FIRST PHYSICISTS

No convincing reason has ever been produced why the inhabitants of the Aegean region in the sixth century BC should have embarked on an intellectual quest which was unparalleled in its own time, and which has shaped the course of western thinking ever since. In this quest, ethics, politics, logic, mathematics, the physical world, and the structure of knowledge were all subject to rational scrutiny in an attempt to establish secure intellectual foundations for human thought and social existence. In scientific terms, the most striking aspect of this intellectual activity was its secularity: fundamental questions about the nature of matter, the mechanism of the cosmos, the geography of the world, or the growth of living organisms were answered in rational terms, without automatic recourse to 'the will of the gods'. In the poems of Homer and Hesiod (eighth century BC), the gods had been portrayed as manipulating the forces of nature and intervening directly in human affairs. In time however, the pantheon of Greek gods came to fulfil a purely cultic function, and ceased to be seen as answering intellectual inquiry. The very shallowness of their religion may have been a key factor in the growth of Greek philosophy: having no profound faith in their gods, their intellect was free to explore realms of thought unknown to other ancient civilizations. The object of their study was *physis* – nature in its widest sense – and they thus became the first the *physikoi* – physicists. The very word which they gave to the world, *kosmos,* meant in its original sense 'order', and this intellectual adventure seems to have sprung from the conviction that the world was an ordered system, which would yield to rational investigation. Noticeable too was the secularity of the scientist-philosophers themselves: in contrast to Babylon and Egypt where science and learning were priestly functions, the Greek scientists were private citizens, teachers for the most part, who had no courtly role or royal patrons to serve. The political freedom of the Greek city states, in contrast to the absolute monarchies of Egypt and Mesopotamia, may not be coincidental, but may have been an important shaping force in intellectual life.

Greek science in its earliest phase was speculative rather than empirical, and from the sixth and fifth centuries BC we have the first names in the history of both science and philosophy: Thales, Anaximander, Heraclitus, Democritus, Empedocles and Pythagoras. The thinking of these men was novel and influential, but there is a serious critical problem in assessing it, for we know it only at secondhand: of their original works, some have survived only in fragments, others only in the reports of later writers. Aristotle considered Thales (*c.*625–*c.*547 BC) to be the founder of Ionian natural philosophy, but he had access to no book written by him. Thales was the subject of a number of scientific stories or legends; the most enduring was that he predicted a solar eclipse in 585 BC, but this is now discounted as no cycle of such eclipses was known, even to the Babylonians, which would have permitted Thales to perform this feat. He was believed by other Greek scientists to have visited Egypt and to have brought back and developed the serious study of mathematics. This too is doubtful, but it is more certain that Thales did address the fundamental question of the nature of matter, the composition of the physical world, which was the kind of intellectual problem that was quite foreign to the Babylonians and Egyptians. Thales's theory, as reported by Aristotle, was that water was the basic element of matter, and that the whole earth actually floated upon water, this being the explanation for earthquakes. Although slightly bizarre, this idea is of interest in its contrast with Hesiod's statement that earthquakes were caused by the god Poseidon. Thales's explanation was a naturalistic, physical one, and while his followers may have disagreed with him, they too sought physical theories to explain the workings of nature. Anaximenes (*fl.*550–540 BC) proposed that the constituent of all matter was air, which undergoes condensation or rarefaction to become all the substances that make up the physical world. His contemporary, Anaximander, considered that this world is only one of a number of possible worlds that have separated from 'the infinite', which is the source of all things, and to which all things will return. Clearly this was intuitive and not experimental, but it represented an attempt to articulate the unity and order that was perceived to be built into matter, and to express it in secular terms, not in the personified terms of the creation myth.

The search for the underlying reality of the universe continued with Heraclitus (*fl.*500 BC), who shared Anaximander's belief in an over-arching law of nature, or *logos,* about which he spoke in a gnomic, almost oriental fashion, as reconciling contraries such as war and peace, heat and cold, life and death. Sometimes Heraclitus spoke of the cosmos as composed of 'ever-living fire', but whether he was proposing fire as a universal element, or merely using fire as an image of flux and transformation is uncertain. This concept of transformation was given its classic expression in the atomism of Democritus (late fifth century BC), in which all matter is built from minute indivisible particles. These are without characteristics, and all the qualities we perceive in different materials are caused by combinations of atoms and the void spaces between them; not simply stone or water, but even categories such as soul and thought are merely arrangements of atoms. The atoms were indestructible, and in constant motion, sometimes slow sometimes swift, as nature's transformations took place, impelled by an inner necessity. On this view the cosmos is a mechanism ruled by an iron determinism, uninfluenced by divinity or mind or purpose.

Where Democritus, Anaximenes or Thales proposed a single fundamental element in the universe, the Sicilian Empedocles (*fl.*450 BC) drew their theories together to formulate one of the most enduring doctrines in scientific history, that of the four elements. Empedocles taught that all material things were composed of earth, air, fire and water, mingled in various proportions and ruled by two cosmic forces – love and strife. From these

The four classical elements. In Aristotle's scheme each element consisted of two paired qualities - fire is hot and dry, water cold and wet, and so on. This scheme was easily represented in a square diagram, and in this medieval example the four cardinal directions and the three continents have been added to compose a world diagram; south is at the top.
Bayerisches Staatsbibliothek, Munich

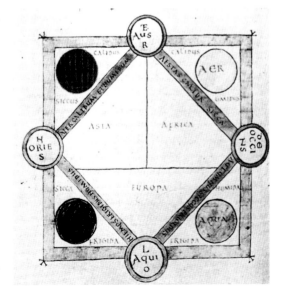

Pythagoras discovering
the mathematics of music,
from a fifteenth-century
manuscript. Legend
recounts that Pythagoras
noticed regular intervals
sounded by blacksmiths'
hammers, then confirmed
this through the sizes or
lengths of bells, strings
and flutes.
The British Library
Add.Ms. 4913 f.17r.

principles sprang 'all things that were and are and shall be, trees and men and women, beasts and birds and fishes, and the long-lived gods too, … for there are these things alone, and running through one another they assume many a shape.' Growth and decay, indeed all processes of change, are merely the re-arrangement of these elements under the direction of the two external forces.

In the thinking of these early 'physicists', it is evident that a vital role was played by immaterial forces – whether these were termed necessity, flux, fire or strife – for having analysed the constituents of matter, the philosopher was inevitably compelled to account for the change, growth and transformation which we find all around us. These creative or organizing forces seem to differ from religious principles only in that they are impersonal. It is in the thought of Pythagoras of Samos (*c*.560–*c*.480 BC) that these quasi-religious principles are made most explicit. Pythagoras founded a brotherhood which was organized very much on religious lines, with strict rules of conduct, and with the aim of penetrating nature's mysteries through mathematical study. Pythagoras travelled as a young man in both Egypt and Babylonia, absorbing their mathematical traditions, and it was undoubtedly in Babylonia that he learned of the relationship that now bears his name (that the square of the hypotenuse of a right triangle is equal to the sum of the square of the other two sides) for it has been found clearly set down in cuneiform texts of this period and before. Pythagoras and his disciples were fascinated by the symmetry of a triangle where the sides were 3, 4 and 5, giving squares of 9, 16 and 25; but disturbed by the case where the short sides are 4 and 5 and the hypotenuse is therefore not a whole number but (in modern notation) 6.4031242 … The reason for this unhappiness was their fanatical search for ratio, proportion and balance in mathematics, which led them to develop a number mysticism which was hugely influential in subsequent ages. They delighted in the discovery of patterns such as the Pythagorean triples (of which the triangular law is one application), or figurate numbers (for example the number 10 formed of 10 points into an equilateral triangle), or 'perfect' numbers, which are the sum of their factors (such as 28, which is $1+2+4+7+14$). It is undoubtedly true that inadequate notation handicapped Greek mathematics in some ways, but it stimulated their use of geometry to demonstrate problems which would later be expressed through algebra, for example equations involving squares would be drawn out as geometric figures:

$$(x + y)^2 = x^2 + 2xy + y^2$$

Cases of numerical proportion in the physical world were especially prized, for example the discovery that the most beautiful musical harmonies were the result of certain well-defined ratios in the length of the string (e.g. in a scale of 12 units, 8 sounds the interval of the fifth, and 6 sounds the octave; as the fifth and the octave were considered harmonious sounds, Pythagoras places 12, 8 and 6 in 'harmonious progression'). This quest for harmony was extended also to the periods of rotation of the heavenly bodies, and their distances from the earth, which it was believed must betray some fundamental pattern. However the reconciliation of the planetary periods and distances with the musical intervals proved an elusive goal, and the 'music of the spheres' was never heard by human ears.

Nevertheless the Pythagoreans were convinced that, since the same harmonies and the same geometric shapes can be produced if certain numerical rules are followed, then those numbers must express the essential order that lay at the heart of nature. In practical terms the activities of the Pythagoreans may not have amounted to a great deal, but their school continued long after their master's death, and gave an immense stimulus to mathematics and even to astronomy. One particularly influential teaching was the belief that the heavenly bodies moved in perfectly circular motion, the circle being the simplest and most perfect shape known to nature. This principle would in time become elaborated into the classical model of the spherical earth within a spherical cosmos.

PLATO AND ARISTOTLE

The Pythagorean school is a paradigm of the scientific impulse to uncover the order and harmony that are believed to be built into the physical universe. But the unity of Greek science and philosophy undoubtedly reached its fullest expression in the thought of Plato (427–347 BC), one of the intellectual giants of history whose influence endured, not least in its Christian form, for more than two millenia, but whose place in the history of science, strictly defined, is ambiguous. The central focus of Plato's thought was the problem of reality and our knowledge of it. Repelled by the materialism of the *physikoi,* Plato was convinced of the omnipresence of design in nature. He expressed this through his celebrated theory of forms: that all the things of this world – material and immaterial – are copies of ideal protypes conceived by the creative power which lies behind the world of nature. Only in this way could Plato explain man's striving after ideals of beauty, courage, justice or love, for the human mind or soul had been endowed at the beginning of time with a knowledge of these forms. The forms are eternal, perfect, unchanging, accessible only to the intellect; while the world of sense experience is imperfect, changing and deceptive. Such a view of nature is potentially hostile to science, carrying as it does the implication that manifold forms of nature are unworthy of study in themselves, and are mere shadows of reality. This impression is softened a little in the single work which Plato devotes to a survey of the natural world, the *Timaeus*, in which the theme of design in the cosmos and the living world is elaborated.

The ordered cosmos presented in the *Timaeus* was created not from nothing, as in the Christian account, but from pre-existing but unformed chaos, by a divinity called by Plato the *demiourgos*, or craftsman. We are told little that is specific about the craftsman, but a great deal about the plans on which he constructed the universe, which was of the simplest and most perfect form – spherical – and Plato proceeds to give one of the earliest accounts of the cosmos as a system of concentric spheres centred on the earth. For Plato all movement is impelled by mind, and regular movement is evidence of a wise and balanced mind. On this view, celestial motion is not caused from outside, but arises from the will of the 'world-soul': the universe is in fact a living creature. As for the specific composition of matter, Plato accepted the existence of the four Empedoclean elements-earth, air, fire and water – but reduced them still further to geometric forms, considering the elements to be composed of the five regular solids (the cube, the pyramid, the octahedron, the dodecahedron and the icosahedron) of classical geometry. Thus for Plato, as for Pythagoras, reality was fundamentally mathematical, and the study of mathematics was a pathway to the highest form of knowledge (although the often-repeated story that the words 'Let no man ignorant of mathematics enter here' stood over Plato's academy, is a legend traceable no further back than the sixth century AD). Man in Plato's view is a divine soul dwelling in a house of flesh, but the body, no less than the cosmos, is designed on the twin principles of reason and necessity. The heart is the seat of the emotions and appetites, while the brain is recognized as the seat of the mind. The function of the blood in feeding the body is supplemented by its role in carrying *pneuma* – the vital force – throughout the body. Physical health is attributed to the balance of the four elements of which the body is composed, while disease results from their imbalance; this was to remain the orthodox theory of disease for two millenia and more.

Timaeus has long been regarded as a dark, problematic work among Plato's oeuvre, for many crucial processes of nature are left unaddressed, and there are many passages where we are not sure if Plato is speaking figuratively or literally, as with the character of the *demiourgos*, or the reality of the geometric atoms of which all matter is said to be composed. The central doctrine of the forms seems to assign to nature the role of a mere shadow or reflection of some higher reality. By inventing a parallel universe of ideal forms, and proclaiming that it is this transcendental realm which should occupy the mind of the philosopher, Plato sounds an almost mystical note. Such a doctrine inevitably throws up powerful barriers to the empirical study of nature's processes, hence Plato's legacy to science might be considered a negative one, were it not for the emphasis on design and on rational creation, which have become absolutely fundamental in western thought.

Although a pupil of Plato, Aristotle (384–322 BC) rejected the metaphysical principles of his teacher, and expounded a realist approach to nature, establishing

Stonehenge, in the water-colour by Constable. Its cultic function may remain forever unknown, but Stonehenge clearly embodied the very precise astronomical knowledge of its Neolithic builders.
Victoria and Albert Museum, London

Hippocrates, appropriately enthroned as the ruler of the Greek medical tradition, from a four-teenth-century Byzantine manuscript.
Bibliothèque Nationale, Paris

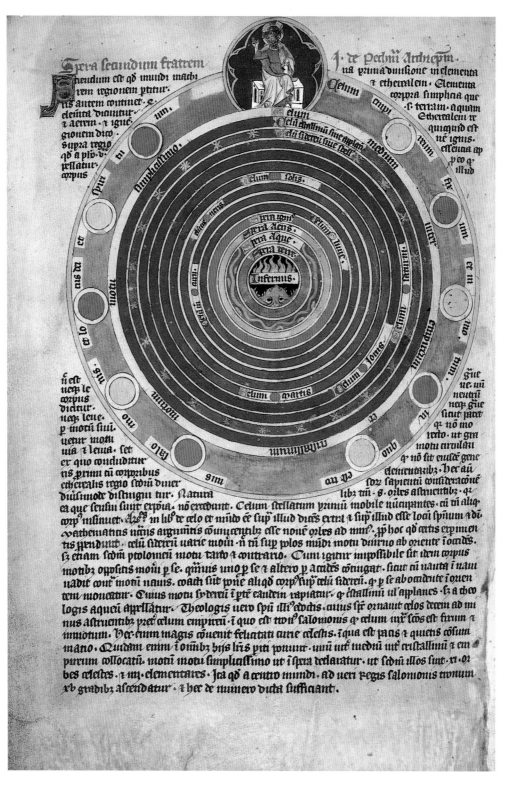

The classical cosmology: the earth, composed of the four elements earth, water, air and fire, is surrounded by the spheres bearing the planets and stars. All surviving illustrations of this classical system are, like this one, from medieval sources.

The British Library Arundel Ms. 83 f.123r.

in the process a scientific language which has never been entirely superseded. The doctrine of forms held no appeal for Aristotle, and he insisted that reality lies in particular things, not in abstract prototypes. Although we can analyse matter and its properties separately, this is a purely epistemological process; in the world of nature, properties are always properties of something. The concept of pure, featureless forms was an intellectual fallacy, and sense experience was the only foundation of knowledge. Aristotle shared Plato's view that the universe is orderly, but explained this not by reference to a designing divinity but by the theory that everything, animate and inanimate, has an individual nature which accounts for its behaviour. Thus it is the seed's nature to grow into a tree, moving from its potential to actual form. This principle was applied also to the inanimate world to create the earliest systematic approach to physics; thus a heavy body falls in order to find its true place – which is at the centre of the universe. This concept seems at first sight a vacuous one, until we relate it to Aristotle's version of the theory of the four elements, in which each pure, unmixed element has a natural place, with earth at the centre of the universe, surrounded by water, above which lie the two lighter elements, air and fire. This ideal system does not occur in nature, for all the elements are mixed together, but it is their striving to find their true level which engenders all the processes of nature. Rejecting Plato's analysis of matter into geometric forms, Aristotle proposed that each element is composed of the fundamental qualities – hot or cold, wet or dry. The diagram of the Aristotelian square displaying the elements and qualities would later become one of the favourite models of medieval science. Plainly, these theories about the nature of the world were built up not from quantitative analysis, but by systematic reasoning, logical deductions from first principles which were supposed to be self-evident – such as that fire and water, heat and cold were such profoundly opposing principles that they must possess utterly distinct identities, and so be classed among the fundamental building-blocks of nature.

The area of Aristotle's physics which was of the most lasting influence was his cosmology. The model of the cosmos as a set of concentric spheres on which the celestial bodies were carried was commonly accepted by Greek thinkers. In the absence of any theory of gravity or inertia, the constituent parts of the cosmos were believed to be linked in some subtle but definitely mechanical system. It was widely known that the earth was spherical, for example from the shadow of the earth which is cast upon the surface of the Moon during lunar eclipses; the geometry of eclipses was understood even before Aristotle's time, and thinkers such as Empedocles had also realized that the Moon shone with reflected sunlight. Perhaps the greatest single puzzle in the movement of the heavens was the difficulty of accounting for the double motion – the daily rotation of the heavens and the much slower positional changes of the Sun, Moon and planets that occur over weeks or months – a puzzle that was ingeniously solved by Eudoxus of Cnidus (*c.*400–347 BC). Eudoxus proposed that the paths of the planets might be explained as the rotation of two, or in some cases three, concentric spheres, set within each other, and whose axes are inclined at different angles. One sphere accounts for each complete daily rotation, while the second and third explain the more limited shifts in position. In particular this theory accounted for the retrograde motion of the planets by showing that the secondary motion sometimes complements the primary motion, and is sometimes opposed to it. In fact Eudoxus calculated that this combined motion formed a figure-of-eight rather than a simple circle. It should be stressed that this was a theoretical and geometric model, and no attempt was made to describe a physical mechanism that could sustain it, but the idea of the cosmos as a nest of concentric, interlocking spheres became fundamental for the next two millenia.

Aristotle accepted the Eudoxan model and elaborated it into a working physical system. Firstly he divided the entire cosmos into two realms, the lower comprising our earth and the lower sky as far as a spherical shell which carried the moon. This realm experiences change and transience, is composed of the four elements, and its motion is linear towards or away from the material centre. The upper realm contains the Sun, Moon and planets, and is unchanging, for 'in the whole range of time past, no change appears to have taken place either in the whole scheme of the outermost heaven or in any of its proper parts'. This led Aristotle to conclude that the physical laws which obtain there are wholly other than those on earth; that the celestial realm is composed of a fifth element, the 'ether', which is pure and unchanging, and that motion in the heavens is not linear but circular. A vital feature of this cosmos is that it forms a continuum: the concept of empty space was quite alien to ancient science. Action at a distance was unthinkable, hence Aristotle's fundamental doctrine of physics – that everything which moves is moved by a constant force. When an object is thrown, it is not momentum or inertia which carries it, but a material vortex in the atmosphere, which, when exhausted, allows the object to fall. In the light of these physical laws, Aristotle described a cosmos in which the celestial spheres were of solid but translucent material, in complex motion one against the other. He saw therefore that it was necessary to introduce intermediate spheres to neutralize or unwind the motion of the major spheres, and designed a celestial mechanism in which the seven planets and the stars were carried on fifty-five interlocking spheres.

The ultimate problem which Aristotle faced was to identify the force which moved the Sun, stars and planets. This he called the Prime Mover, and seems to have been unable to conceive of it other than as a personal deity, which interacts with the spheres not by forcibly impelling them, but by drawing them towards himself with a transcendent power. This theory was to inspire Christian theologians to identify the Prime Mover with God. Yet there were a number of serious difficulties here, since Aristotle proposed not a single Prime Mover for the whole universe, but one for each celestial sphere, and he stated explicitly that the universe can have had no beginning, since nothing can be created out of nothing. In medieval thought, the forces moving the celestial spheres were transformed into angels.

This area of Aristotle's thought was of paramount importance, dominating western astronomy for almost two thousand years. Yet in a second field, that of biology, his unrivalled originality was far ahead of its time and was long neglected. He carried out detailed first-hand studies of hundreds of animal species and devised major classifications such as blooded (i.e. red-blooded) and bloodless, viviparous and oviparous, vertebrate and invertebrate, and so on. Although he could offer no explanation for the multiplicity of life-forms, he did recognize that they are interrelated, that some are higher than others, and, believing that all living things shared a vital principle or soul, proposed a scheme of classification rising from the vegetative soul through the animal soul to the rational soul. This vitalism was in direct contrast to the atomism of Democritus, which held that life was capable of a mechanistic analysis, and it was one area of Aristotle's thought which fitted well with the Neoplatonic concept of interrelated levels of being, so that later, in the middle ages, various diagrams of the ladder of nature became, like the square of the elements, familiar teaching models of medieval science. The soul or vital principle gave form or actuality to the material body, which was possessed only of potentiality. In the field of human physiology, Aristotle taught that the female provided the material base of life in the form of menstrual blood, while the male semen bore the form or soul. Although he did not dissect the human body, he carried out many animal dissections and drew lessons by analogy, for example in his embryological work he saw the beating heart as the first sign of life, and observed the blood vessels radiating from it. From this he concluded that the heart was 'the first to live and the last to die' – that it was the very centre of the life system, and that its organic function was to produce blood. The existence of the nervous system escaped his notice, and he considered the brain to fulfil a secondary cooling function after that performed by the lungs.

It cannot be said that Aristotle identified a final or controlling power in the life sciences as he did in cosmology. What principle is at work in the diversification of life-forms or in the evident design of the body, he does not explain. The form or soul of the whole living creature determines the structure and functioning of each of its parts, but no divine or metaphysical power is invoked to direct nature from the outside, as Plato's *demiourgos* does. As a systematic philosopher of nature, Aristotle had no equal in the ancient world, producing a corpus of teaching on the physical and biological sciences that would retain its authority for two thousand years. A determined realist, he rejected the metaphysics of Plato and reinstated natural forms and forces as the fundamental field of study. He was often guided by *a priori* assumptions, yet he aimed always at quantification and rational description. Many of his theories have inevitably been superseded, but it was he who established physics and biology as subjects of analytical study, and he who fashioned a language capable of handling new and fundamental concepts: when we use words such as energy, potential, quantity, essential, causation, category or substance, we are still treading the borderland between science and philosophy marked out by Aristotle.

The influence of Plato and Aristotle was disseminated through their schools in Athens, which survived for approximately two centuries after their deaths; but there were other schools with a diversity of philosophical and scientific approaches. One of the most famous, and slightly antedating Aristotle, was the medical school on the island of Kos that is forever associated with the name Hippocrates. Hippocrates (*c.*460–370 BC?) is a profoundly obscure figure in classical history, effectively no more than a name that has been attached to some sixty treatises which mark the beginning of critical, secular medicine; whether an individual named Hippocrates composed them is uncertain. Traditional Greek medicine centred on the cult of Asclepius, who is mentioned as a historical figure in the *Iliad,* but who was worshipped as the god of healing by the sixth century BC. It was reputed that Asclepius visited and healed the sick during their dreams, and in temples dedicated in his name physicians practised a strongly magical form of medicine. The snake-entwined staff which is the symbol of Asclepius suggests the practice of sympathetic magic – the overcoming of poison through its controlled presence. The Hippocratic treatises are the first written medical texts, composed in the Ionic dialect, which thus became the language of Greek medical science. In these writings, methods of healing are taken out of the realm of magic and folklore, and the agency of the gods is discounted in favour of a theory of disease as a malfunction of the body-systems. For example a whole class of disorders including epilepsy were popularly termed 'the sacred disease', but the Hippocratic author protests that

'Those who first called this disease sacred were the sort of people we now call witch-doctors, faith-healers, quack and charlatans … by invoking a divine element they were able to screen their own failure to give suitable treatment, and called this malady 'sacred' to conceal their ignorance of its nature'.

In Hippocratic thought, the underlying cause of disease was considered to be an imbalance in the four fluids or humours that make up the body – blood, phlegm, yellow bile and black bile. The origin of this doctrine is obscure, but it developed a tenacious hold on the scientific mind, and became elaborately connected with the four elements, the four seasons, with planetary deities and so on. But the distinction of the Hippocratic approach was that this general theory was less important than a comprehensive and precise study of symptoms, in which the whole body was to be scrutinized in an effort to build up a complete picture of the patient and his ailment. Treatment was largely dietary and herbal, while wounds and tumours called for surgery or cauterization: 'What drugs fail to cure, that the knife cures; what the knife cures not, that the fire cures; but what the fire fails to cure, this must be called incurable.' Mercifully, natural anaesthetics were available in the form of opium, mandragora, and of course wine. Perhaps surprisingly, the anatomy and physiology of the Hippocratic school was distinctly sparse and unsystematic; the difference between veins and arteries was unknown, and the nervous system was confused with ligaments. As far as we know there was no official dissection of the human body, and analogy with dissected animals must have given rise to many errors. Yet the feature which was of permanent value was the high professional standards demanded of practitioners, in a field where magic and charlatanism had prevailed. However, even the famous oath presents historical problems, for it appears to contradict on a number of points the Hippocratic writings, for example in forbidding abortion, a practice whose technique the writings openly discuss. It seems likely that the present form of the oath evolved over a long period of time, thus accounting for these discrepancies. The major innovation of Hippocratic medicine was undoubtedly the view that the body was an organic whole, a natural unit in a natural world – a discovery which continues the theme of the secularization of science in Greek civilization.

HELLENISTIC SCIENCE

With the conquests of Alexander the Great, a process of cultural interaction was set up in which Greek thought was carried throughout the ancient near east, while Greece in turn was opened to intellectual forces from Egypt, Persia and India. Athens ceased to be the intellectual centre of this world, and Alexandria, with its great library and 'museum' (something like a research centre dedicated to the muses, the goddesses of the arts and sciences) became the focus of research in Hellenistic science from around 300 BC until the third century AD. Simultaneously the worlds of philosophy and science began to diverge. After Aristotle, philosophy became the art of living, with ethics at its heart, and the pursuit of science defined itself more precisely as the mastery of mathematics, biology, astronomy or mechanics. The philosophical attitude to science is summed up in Epicurus's dictum that science 'has no other goal but tranquillity ... for if we were not disturbed by our fears of celestial things and of death ... we should have no further need of natural science.' Plato and Aristotle had disagreed as to whether or not nature was fundamentally mathematical. The most famous mathematician in history, Euclid, did not advance that debate in strictly philosophical terms, but he crystallized a mathematical language which certainly suggested that it was. Like Hippocrates, Euclid is no more than a name, an impersonal authority, who reputedly taught in Alexandria around 300 BC, and whose *Elements* became the most enduring and widely-studied secular book in the western world, dominating the teaching of mathematics for more than two thousand years. Principally

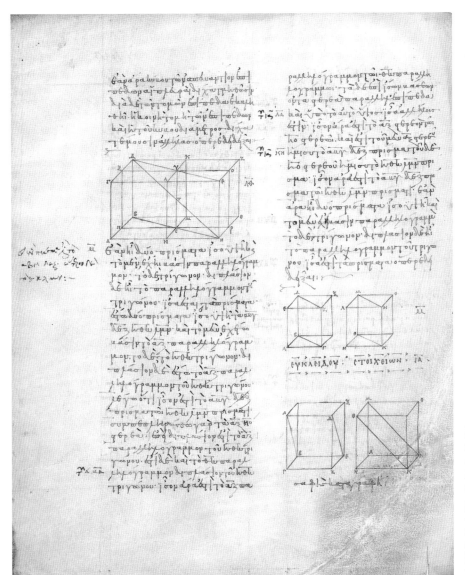

Euclid's geometry: a
ninth-century manuscript
of Book XI of the
Elements showing the
three-dimensional
geometry of prisms.
For two thousand years
the *Elements* was the most
studied non-religious text
in the western world.
Vatican Library, Rome

devoted to plane and solid geometry, Euclid's work sets out scores of theorems
and problems in the classic form of definition, demonstration and proof, which
came to be seen as a paradigm of logical thought. Euclidean space, composed of
points, lines, angles and circles, was infinite but homogeneous. Certain classical
problems proved insoluble, such as doubling the cube or squaring the circle, but
Euclid fashioned a language and method of mathematics capable of application to
astronomy, optics, mechanics and engineering.

Archimedes (*c.*287–212 BC) was celebrated as a mechanical engineer, reportedly building ballistic machines, water-raising screws, compound pulleys,
working models of the solar system, celestial globes, and, more doubtfully, a huge
concave mirror powerful enough to focus the sun's rays and set fire to ships at sea.
These practical achievements however formed no part of Archimedes's written
works, which are all of a theoretical nature, with the exception of his work on

hydrostatics, which includes Archimedes's principle, stating that a solid immersed in a fluid will be lighter by the weight of water which it displaces. He calculated that the surface area of any sphere is four times that of its greatest circle, while the volume of a sphere is two-thirds that of the cylinder in which it is inscribed. His pride in these discoveries was shown by his decision to carve a sphere and a cylinder upon his tomb; this was verified by Cicero when he uncovered the grave two centuries later. He worked intensely at squaring the circle by the Euclidean method of exhaustion, which proceeds by inscribing within the circle regular polygons whose area is calculable; by this method he arrived at a value for π of between $3^{1}/_{7}$ and $3^{10}/_{71}$. He was able to reduce the mechanics of levers to pure geometry, and to show that magnitudes balance at distances from the fulcrum in inverse ratio to their weights; the balance beam became a line, the weights are points positioned in strict proportion upon it.

The taste for mechanical devices like those of Archimedes was carried further over many years by Alexandrian scientists, from Ctesibius (*fl.c.*270 BC) to Hero (*fl.c.*AD 60). These men invented devices such as water-clocks, and automata powered by controlled weights, compressed air, heat and even steam. Trick jars that

Alexandrian mechanisms described by Hero in the first century AD. *Left*: steam from the cauldron rises into the sphere, causing it to rotate like a turbine. *Right*: heat from the fire pushes hot air down into a tank of oil, and the oil is forced up through the figures to drip onto the fire in the manner of an altar libation.
By courtesy of Oxford University Press

poured out either wine or water, model animals that drank when they were offered water, model ships that were wrecked while dolphins leapt around them – all these displayed an impressive mastery of hydraulics and thermodynamics, raising the interesting question why these skills were not channelled into more practical or permanent technologies. One of the most striking applications of simple Alexandrian mathematics lay in the success with which the astronomer Eratosthenes calculated the circumference of the earth around the year 235 BC. Observing a vertical column at Aswan in Upper Egypt (which lies on the Tropic of Cancer) at noon on the summer solstice, he noted that it cast no shadow. Some 500 miles north in Alexandria at the same moment, the shadow of a similar column formed an angle of seven degrees to the vertical. Eratosthenes reasoned therefore that 500 miles was $^{7}/_{360}$, or $^{1}/_{51}$, of the circumference of the globe. The result, approximately 25,000 miles is within 6% of the true value. In the realm of pure mathematics, Apollonius of Perga (*c.*262–*c.*190 BC) established the fundamentals of conic sections, introducing the terms parabola, hyperbola and ellipse for the different intersections of plane and cone, while in astronomy it was he

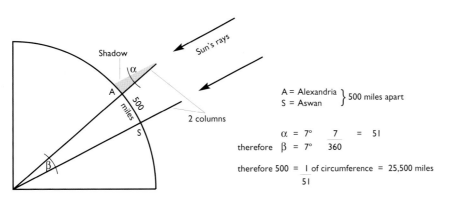

A = Alexandria
S = Aswan } 500 miles apart

$\alpha = 7°$ $\frac{7}{360} = 51$

therefore $\beta = 7°$

therefore $500 = \frac{1}{51}$ of circumference $= 25,500$ miles

who replaced the Eudoxan planetary theory with the superior cycle-and-epicycle model. In Apollonius's scheme, the main planetary sphere carries on it smaller spheres revolving on their own axes, thus also explaining changes in the speed of the planets, and their retrograde motion. This model would be enshrined in the work of Ptolemy and become standard until Copernicus.

Geometry was combined with natural philosophy in one of the characteristic sciences of the classical world – optics. As the primary means of interraction between man and his environment, optics held a special importance for philosophers who were attempting to define the sources and validity of knowledge. All Greek thinkers seem to have conceived of vision as a material process, in which a subtle, but definitely physical, substance passed between the eye and the perceived objects. Perhaps the most surprising aspect of this approach was the belief that this substance issued from the eye; thus Plato describes a 'fire' radiating from the eye, along which impressions of objects are conveyed to the soul. To the obvious problem of how darkness can exist in this scheme, Plato answers that the optical fire is activated only when it coalesces with sunlight. Reversing this process, Epicurus stated that all objects constantly emit rays which strike the eyes, rays which contain replicas of the parent object which speed through space to the observer's eye. As they travel they become blunted and rounded by collision with the air, thus explaining why distant objects lose definition. It was Euclid who devised the first geometric theory of optics by suggesting that the rays emerging from the eye form a visual cone, whose apex lies in the eye and whose base becomes ever broader as one observes distant objects. Euclid used this construct to analyse perspective, concluding that distant objects appear higher in a perspective view because they intercept a higher ray in the visual cone:

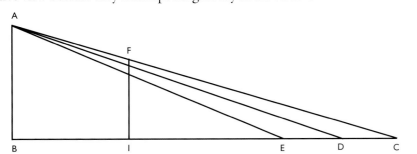

Ptolemy (see below) carried this form of analysis much further, investigating the geometry of reflection and refraction. Using a mirror and a disc graduated into degrees and equipped with sighting tubes, Ptolemy discovered the fundamental

Eratosthenes calculated the circumference of the earth with remarkable accuracy in the second century BC by measuring one fiftieth of an arc of a great circle, between Aswan in upper Egypt and Alexandria. A column cast no shadow at the summer solstice in Aswan, on the tropic of Cancer, while 500 miles north at the same moment, the shadow of a similar column formed an angle of 7 degrees. 500 was therefore $^7/_{360}$ of the circumference of the earth – approximately 25,000 miles.

Euclid's geometry of vision. A is the observer's eye: the points E, D and C are progressively more distant and they invariably appear higher than the foreground because rays of light from them occupy higher positions within the visual cone; this is clearly seen when they intersect the plane FI.
By courtesy of D.C. Lindberg

rule that the angle of incidence is equal to the angle of reflection, and he went on to show that our image of a reflected object has a definable position behind the mirror. Experimenting with concave and convex mirrors he showed that their distortions are applications of optical laws, as were the effects of refraction in water for example. This dual approach to optics – physical and geometric – established a scientific discipline which flourished for centuries and which, most unusually, combined experiment with elaborate theorizing, in pursuit of that most elusive problem, the nature of light.

Alexandria also possessed a school of medicine which owed at least some of its achievements to the fact that dissection of the human body was not taboo in that city, as it was virtually everywhere else in the Hellenistic world. Exactly why this was we do not know, but the practice was common by the third century BC, and it enabled physicians such as Herophilus and Erasistratus to make considerable advances, although reports of their methods are macabre:

> ' Herophilus and Erasistratus were given criminals from the prison by the kings of Egypt, and dissected them alive: while they were still breathing they observed the parts which nature had formerly concealed, and examined their position, colour, shape, size, arrangement, hardness, softness, smoothness, connection… Nor is it cruel, as most people allege, by causing pain to guilty men – and only a few of them – to seek out remedies for innocent people of all time.'

Herophilus (*fl.c.*270 BC) made important anatomical discoveries concerning the brain and nervous system, the eye and the vascular system. He succeeded in tracing the connections of major nerves to the spinal column and the brain, and concluded that it was the brain and not the heart which was the focus of sensation and the seat of the intellect. His name is still preserved in certain areas of the brain. He considered that the nerves functioned as pathways for a vital force which he, like other Greek thinkers, termed *pneuma* – breath or spirit. He discovered that the optic nerve was hollow, and therefore especially suitable for carrying an aesthetic *pneuma*. He recognized the pulse as an index of the heartbeat and expounded on its use as a diagnostic aid. He was deeply interested in gynaecology, and accurately described the ovaries, uterus and cervix. This keenly anatomical approach was shared by Erasistratus (*fl.c.*260 BC), but he was more willing than Herophilus to use his findings as the basis for physiological theories. From his dissections of humans and animals he correctly deduced that the convolutions of the brain were an index of intellectual development. He believed that food was absorbed from the stomach and used by the liver to form blood, which was then conveyed by the veins all parts of the body. The heart's function was to distribute *pneuma,* which was inhaled by the lungs, throughout the body via the arteries. The nerves he considered to be carriers of a form of psychic *pneuma,* which was refined within the brain, which he too identified as the seat of the mind. He believed that most disease sprang from an excess of blood, caused by over-eating, hence his preferred therapies were dieting and blood-letting. The findings of the Alexandrian school were resisted by physicians elsewhere who had no experience of dissection, and who branded these new theories as over-intellectualized. Many of Erasistratus's inferences were inevitably mistaken, and functions which could not be seen were explained by reference to invisible forces such as the *pneuma,* but given the complexity of the nervous system, this was scarcely unreasonable. The Alexandrians' keen, empirical study of the human body, and their insight that it functioned through a number of discreet but interlocking systems, gives them a special place in the history of scientific medicine.

Another later and highly influential figure in Hellenistic medicine who studied in Alexandria was Dioscorides (*fl.*AD 50–70), whose work on pharmacology bequeathed a permanent legacy to late classical, Islamic and western science. Principally associated with the medical uses of plants, his great work *De Materia Medica*, also lists the properties of many animal products and minerals. His approach was eminently practical, and he frowned on any philosophical attempts to account for the effects of his drugs. Among the substances accurately described by Dioscorides and still found in any pharmacopoeia are aloes, belladonna, calamine, cinnamon, galls, ginger, marjoram, mercury, opium, sulphur, thyme, wormwood and many others.

ASTRONOMY: THE LEGACY OF PTOLEMY

The field in which Hellenistic science reached the greatest maturity was astronomy. In the five hundred years after the work of Aristotle, spherical geometry was employed in two major areas – constructing models of the cosmos, and mapping the stars. The image of the universe described by Plato and Aristotle became universal in Greek thought: that the spherical earth lay at the centre of a nest of concentric spheres which carried the Sun, Moon, planets and stars in their paths around the unmoving earth. The stars were fixed in relation to each other in the outermost sphere, and revolved *en masse* about a polar point in the northern sky (and in the southern sky too, although it was invisible). In the hands of Eudoxus and later theorists, this general picture became more sharply focused into the conceptual model of the celestial sphere, with an axis and two poles, and the earth as its centre point. Upon this sphere, Greek skills in spherical geometry permitted the tracing of a structural frame of reference which enabled positions to be plotted by coordinates measured from notional base-lines, analogous to latitude and longitude. The north-south positions (later known as declinations) were measured from the ecliptic, the apparent path of the sun through the course of the year; while the longitudinal positions (later known as right ascensions) were taken from the vernal equinox, the point at which the ecliptic crosses the celestial equator. When these key points were joined, a model of the celestial sphere could be drawn, or actually constructed as the instrument we know as the armillary sphere, which was known and used in Hellenistic science, and later became the visual symbol of astronomy. It is in fact an outline globe, and celestial globes are also known to have been made by Hellenistic astronomers. Within the main structural bands of the armillary, additional bands could be added to represent the spheres of the planets.

How were the stars mapped or displayed? Like all sky-watchers, the Greeks needed to impose order on the myriad of stars in the night sky, and by the fifth century BC had taken over a number of constellations from Babylonian astronomy, and had added some from their own mythology. A full description of the classical constellations was written by Eudoxus around 380 BC, and these star groups would endure to the present day, serving the astronomer as notional sky-zones. Eudoxus's own text is lost, but it is known through a work based extensively on it, the *Phaenomena* by Aratus of Soli (*c.*310–240 BC) which became the standard guide to the constellations in the Hellenistic world, and was translated into Latin by Cicero. The constellations and the division of the circle into 360 degrees are two of the most obvious borrowings from Babylonian astronomy, but precisely when and how these ideas were brought to Greece we cannot say. There existed in Greek science by the fourth century BC a clear geometric model of the cosmos, and a functional map of the sky. With these tools, Greek

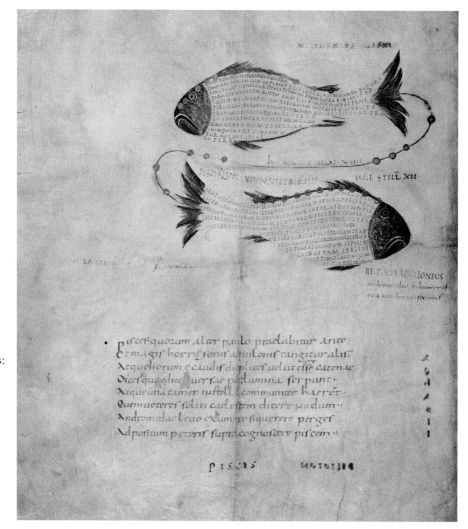

The classical constellations: the words of the third-century BC poem *Phaenomena* by Aratus of Soli are shaped into each of the constellations described. This tenth-century manuscript uses the Latin translation by Cicero.

The British Library
Harley Ms. 647 f.3v

astronomers set out to chart and to explain the movements of the heavenly bodies, with the ever-present presupposition of uniform circular motion, for this was the perfect form of movement and therefore appropriate to celestial objects. This process culminated in the work of Ptolemy of Alexandria (*fl.*AD 130–170), a towering genius of Hellenistic science, whose very pre-eminence means that we now know little of his predecessors, for he summed up and surpassed their achievements so completely that only fragments and second-hand reports of their work have survived. A particular fame however is attached to one of them, Aristarchus of Samos (*fl.*280 BC), for his suggestion that the Sun and not the earth was the centre of the cosmos. Unluckily, we know of this theory only through a brief report by Archimedes, and are ignorant of the reasoning behind it. It clearly accounted well for the observed motions in the heavens, but in proposing a moving earth, Aristarchus was violating all accepted wisdom, and indeed the evidence of our own senses. Aristarchus's position attracted little attention in its own time, but it was never entirely forgotten – Copernicus was certainly aware of it eighteen centuries later – and it is a striking example of the creativity of Greek science.

Ptolemy's astronomy is embodied in his book, the *Almagest*, invariably known by this, its Arabic title, which means simply 'the Greatest'; its original Greek title was 'The Mathematical Compilation'. He begins with a philosophical justification for studying astronomy: the study of physics and the things of this world is concerned with changeable, corruptible elements, while astronomy is on a higher plane, turning the mind to the divine order in the universe. He outlines the methods of spherical geometry which he will employ in describing celestial movements, and then proceeds to construct a mathematical model of the paths of the Sun, Moon and five planets. His concern is always to reconcile the observed positions of these bodies with the principle of uniform circular motion, and to do this he used a series of observations of his own, but also many others recorded by earlier astronomers, always of course with naked-eye instruments. Since, as we now know, the motions of the heavenly bodies are not uniformly circular, and the structure of the cosmos was not as Ptolemy believed it to be, he was forced to employ enormous ingenuity to work out his geometric models. He used two principal devices: first he accepted and enlarged Apollonius's theory that the heavenly bodies possessed a double motion – a large sphere (the 'deferent') which carried upon in itself the axis of a smaller sphere (the 'epicycle'); and second he proposed that the large sphere need not be centred precisely on the earth, but was eccentric to it. The earth was still the central body of the cosmos, but the celestial spheres actually revolved around an 'equant point' slightly distanced from the earth. With these devices Ptolemy produced a scheme of the heavens which employed sophisticated geometric models to enable any planetary or stellar position to be predicted. In nearly fourteen centuries following Ptolemy, this system was sometimes criticized as over-complex, but no convincing alternative ever displaced it from its dominant position: it was the cosmic system accepted by pagan, Islamic and Christian theology as evidence of supreme cosmic design. It is a purely geometric system, and what physical forces sustained it Ptolemy does not describe. He accepted the Aristotelian view that the cosmos is finite and that it is a continuum with no empty space, so that one celestial sphere is directly tangent to the next; but what these spheres really are, or what moves them, we never discover.

The *Almagest* also contains the most complete star catalogue from the ancient world. Ptolemy lists 1022 of the brightest stars, grouped into 48 constellations. Only a handful of the stars are named, for most of the names which we now employ are of Arabic origin, but instead Ptolemy describes their position in each constellation, thus Aldebaran becomes star number 14 of Taurus, 'the bright star of the Hyades, the reddish one of the southern eye of the Bull'. He gives each star its celestial latitude and longitude, and an estimate of magnitude in a scale of six, enabling the reader to construct a star globe if he wishes. The stars in Ptolemy's catalogue and his 48 constellations constituted the map of the heavens until the early seventeenth century, indeed it has been said that from the second century to the sixteenth, all astronomy was but a commentary on Ptolemy. In one significant area Ptolemy acknowledged his debt to his most important predecessor, Hipparchus of Rhodes (*fl.*150–125 BC), who had noted through careful observation

Geometry of the planetary orbits: the Ptolemaic system, in which each planet is carried on two cycles (or sometimes more) with the smaller epicycle centred on the main oribital cycle, is seen in this fifteenth-century manuscript.
Pierpont Morgan Library, New York

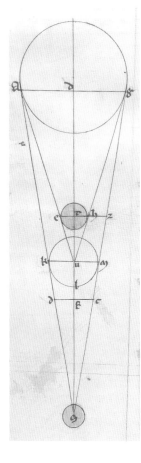

Calculating cosmic
dimensions: this picture
from a manuscript of
Ptolemy's *Almagest* is not
intended to explain
eclipses, but uses the
shadow cones to derive
the distances of the Sun
and Moon; Ptolemy's
estimate however was not
close to the true figure.

The British Library Burney Ms.
275.f.459v

that, although the stars were indeed fixed, the entire starry sphere is very slowly rotating from east to west, so slowly as to amount, in Hipparchus's view, to about one degree every hundred years. Now known as the precession of the equinoxes, it is in fact approximately one degree in 72 years, and its effect is a gradual change in the longitudinal coordinate of any star over a period of time. Through the centuries in which it was used, the star catalogue of the *Almagest* would have its longitudes constantly revised.

Ptolemy's achievement did not end with his astronomical work, for he was equally definitive and scientific in his geography. Here too he proposed a geometric model – the sphere – divisible mathematically by coordinates, and he compiled a gazetteer of some 8,000 places in the world, each with their latitude and longitude. He explains how to project a map of the spherical earth onto a two-dimensional surface, a task admittedly somewhat easier for him, for his world extended only through Europe, most of Asia and northern Africa – approximately one quarter of the globe. He accepted that there might be other continents and other peoples in the unknown portions of the earth, but he did not speculate on their nature. His geography had less influence than his astronomy, for it was lost to the west for more than one thousand years, although preserved by Islamic scientists, until it was revived during the Renaissance.

Despite all the ingenuity they displayed, the geometric systems of Hellenistic astronomy remained technical and qualitative, failing to place man satisfactorily within the cosmos. In order to achieve this, a further physical dimension and a psychological framework were required. In two of his further works Ptolemy is an excellent guide to certain other philosophical concerns of the Hellenistic world: the scale of the universe, which would locate man and his world, and the power of astrology, which would explain the mutual dependence of the cosmic elements.

Within the enclosed universe of Aristotle and Ptolemy, the Moon was always known to be nearest to the earth because it eclipsed all other celestial bodies. The stars were the fixed background, hence they must be farthest from the earth. For the other bodies, it was logical, in an earth-centred system, that speed was an index of their distance, those moving slowest being farthest from the earth. Saturn, Jupiter and Mars were thus placed in descending order, but between the Sun, Mercury and Venus it proved impossible to distinguish. In his short work *The Planetary Hypothesis,* Ptolemy gives a physical scale to the abstract geometry of the *Almagest,* and using calculations based on parallax, on the angular size of the Sun and Moon, and on eclipse data, he arrived at some fundamental dimensions of the cosmos, which he expressed in multiples of earth radii, but which equate in modern terms to these: from the earth to the Moon is 160,000 miles (true figure 240,000); from the earth to the Sun is 3,100,000 miles (true figure 93,000,000). He calculated that the Sun was some five times the diameter of the earth, while the Moon was a quarter the size. He was aware that Jupiter and Saturn were considerably larger than the earth, while Mercury and Venus were smaller; Mars he considered to be slightly larger. From the earth to the stars was 50,000,000 miles. Since the stars were conceived to be set in a single sphere at a fixed distance from the earth, the total diameter of the universe was thus believed to be 100,000,000 miles. Although this is a large figure, and no doubt more so to Ptolemy and his contemporaries, it proved that the universe was a finite, enclosed system, well within the bounds of comprehension. This picture of the measured universe was refined, especially by Islamic astronomers, during the following centuries, but not seriously questioned until the Copernican revolution.

The final component of Ptolemy's legacy to both the Islamic world and the west was, perhaps surprisingly, astrology, for his work *Tetrabiblos* became an authoritative source-book in the field. On the analogy of the physical effects of the Sun and Moon upon the earth and its inhabitants, Ptolemy argues that the other heavenly bodies must also exert their own, more subtle influences on human life. Accepting the universal Greek belief that the stars and planets are divinities, he regards it as logically and theologically obvious that the heavenly bodies shape the course of human life. Faced with the fundamental problem as to how these astral influences actually work, Ptolemy argues that they are physical forces emanating from the stars and planets, as do light or heat. The relevance of the geometric models of the *Almagest,* is that mathematical astronomy allows the astrologer to chart celestial positions, and therefore to predict these complex influences, both on individuals and on the world as a whole, through natural and historical events. Once again, the importance of the finite, measured universe is to provide an enclosed system where these forces can operate; if the universe were infinite, these forces would dissipate themselves unpredictably in empty space. Ptolemy's thought thus proceeds from austere, mathematical science to a philosophical concept of a harmonious universe where nature is the setting and men are the actors in a complex but comprehensible drama. Was such a picture of the universe scientific or religious, or was there no difference?

SCIENTIFIC THOUGHT IN THE ROMAN ERA

From its foundation in the sixth century BC, Rome grew in power until by 260 BC it controlled the whole of Italy, and two centuries later it had become the master of the entire Mediterranean region. Roman civilization was centralized, militaristic, pragmatic, masculine; learning was regarded as purely utilitarian, while the investigation of problems in philosophy or natural science was the leisure pursuit of a few scholars. Roman engineers were rightly famous for their massive building achievements, which formed the essential infrastructure of their imperial power, but as far as we can tell theoretical science played little part in them. There was no mathematical body of knowledge concerning weights and stresses or strengths of materials, even in a writer like Vitruvius, and Roman arches were always limited to the semicircle, never employing ellipses. They knew the use of toothed gears and pulleys for raising weights, and they used pumps, but again without theoretical mechanics, while their considerable metallurgical skills must have been transmitted as an oral craft. There seems to have existed a fundamental difference between the Roman mind and that of their Greek predecessors, which explains the paucity of original scientific thought among Roman scholars, whose works tended merely to condense and simplify the body of learning inherited from Greece. During the Hellenistic period, Rome was open to the Greek intellectual tradition, and the basic tenets of astronomy, medicine, mathematics and geography were absorbed from Greek scholars living in Italy and employed as teachers in many Roman households. But Roman thinkers were unable to enrich this tradition: like the Greek genius for critical philosophy, the Roman distaste for abstract thought is simply an insuperable fact of history.

Typical of the flow of Greek knowledge into Rome was Posidonius (*fl.*135–51 BC), head of the Stoic philosophical school of Rhodes, who visited Rome and numbered men such as Cicero among his pupils. Posidonius wrote histories analysing the rise of Rome, and sketched a theory of learning in which he argued that philosophy deals with fundamental principles while natural science tackles specific problems. He taught that there were three great causal powers, God,

Nature and Fate, and that man's power to manipulate or appease them was tiny. He was nevertheless keenly interested in geology, geography and astronomy, and believed that the configurations of the heavens could reveal the future. Cicero absorbed these lessons, and became concerned to convey Greek wisdom to the Roman world. He translated into Latin certain basic scientific works such as Plato's *Timaeus* and Aratus's *Phaenomena,* and composed many dialogues in which he expounded a form of Platonism, emphasizing the analogy between man's animation by a soul and the universe's direction by God. This microcosm-macrocosm concept was highly influential in medieval Christianity, as was the vision of the celestial spheres described in his work *The Dream of Scipio.*

Perhaps the best-known document of Roman science, though hardly the most typical, is the poem *On the Nature of Things* by Lucretius (*c.*95–55 BC). Lucretius explains all phenomena in the universe by an extreme form of atomism, drawn from that of Democritus and Epicurus. In contrast to the finite and ordered world of both Plato and Aristotle, Lucretius's universe is unlimited in space and may contain a plurality of cosmic systems, of which ours is but one. The universe is composed of minute, property-less atoms, moving in a void without external control, whose random motions create all the compound substances out of which all things, living and inanimate, have built themselves. The human soul itself is such a substance, which dissolves at death. There is no design and no deity in this universe, and the only distinction between living and non-living things lies in the way the atoms combine. The weakness of this viewpoint is its failure to suggest any laws governing this atomic combination, or more precisely its conviction that no such laws exist. Lucretius undoubtedly represented a minority view, for his was a forbidding universe, hostile to man's reason and instinct for design. In the Christian era it was, not surprisingly, attacked for its atheism, and was virtually forgotten, but since its effective rediscovery in the sixteenth century, it has attracted much interest as an anticipation of atomic theory.

Such philosophical approaches to natural philosophy as that of Lucretius became less and less common in Roman thought, whose scholars were more attracted by encyclopedic summaries of facts and leading ideas. The archetypal figure in this approach to science is Pliny the Elder (AD 23–79), whose vast work, *Natural History,* covers geography and astronomy, zoology and botany, anthropology and history. We are told by Pliny himself that the greater part of his book is distilled from earlier texts, not from personal observation, and its best-known features are the uncritical reports of monsters and natural marvels, which so impressed themselves on the European imagination that they were repeated for fifteen hundred years. Pliny may have believed that he lived in an age of intellectual decline, for his reiterated concern is to preserve the knowledge of the past and transmit it to the future. His industry was enormous, and he reportedly worked day and night on his researches, justifying his motto that 'to live is to be awake'. His passion for facts is said to have led to his death, for it was while observing the great eruption of Vesuvius at close quarters that he met his end.

Roman interest in the exact, mathematical sciences was at a rather elementary level. Euclid's geometry for example did not appear in a Latin translation until that by Boethius in the sixth century AD. Astronomy was represented by the Latin versions of Aratus's *Phaenomena,* with its descriptions of the constellations. The astrological component in astronomy was developed for the Roman audience by Marcus Manilius, whose poem, known as *Astronomica,* became the standard guide to the subject. One of the great texts of applied science for which there was no superior Greek model was Vitruvius's *On Architecture,* composed around 30 BC. It treats subjects as diverse as building materials, the astronomy of sundials,

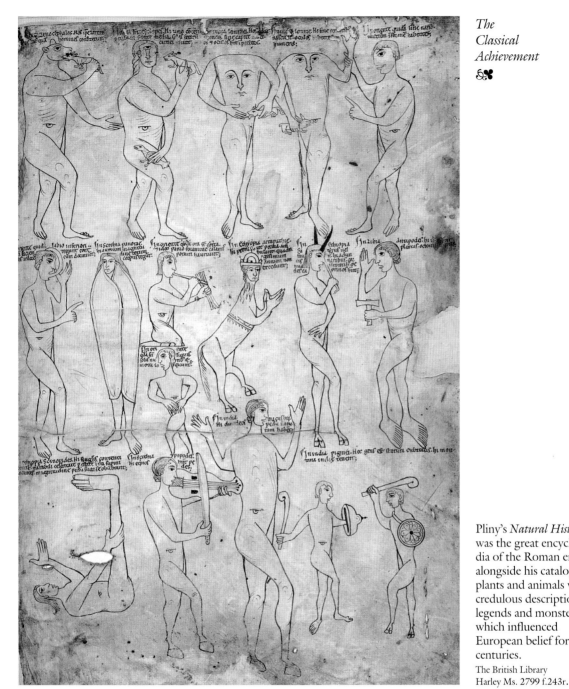

Pliny's *Natural History* was the great encyclopedia of the Roman era, but alongside his catalogue of plants and animals were credulous descriptions of legends and monsters which influenced European belief for centuries.

The British Library
Harley Ms. 2799 f.243r.

the making of water-clocks, bridge-building, military siege-engines and an analysis of the classical orders of architecture. Vitruvius's work is a rich source of Roman technological history, and it exercised great influence when it was rediscovered in the Italian Renaissance.

Even in the eminently practical art of medicine the Romans were dependent on Greek models and expertise. The medicine of the late classical period is overshadowed by the figure of Galen (AD 128–201), a native of Pergamum whose prolific works were written in Greek, the language of medicine since Hippocrates, but who spent much of his life practising and teaching in Rome. Galen extended the theory of the four humours, the four constituents of the human body, by arguing that each individual is dominated by one of them, thus producing the four classical temperaments: blood engenders the sanguine temperament, phlegm the phlegmatic, yellow bile the choleric, and black bile the melancholic. These humours are not to be understood as visible systems like the blood, but as constituents of the very tissues, and as related to the four elements: blood was the counterpart of air, phlegm of water, yellow bile of fire, and black bile of earth. Compared with his Alexandrian predecessors, Galen had little if any opportunity for human dissection, for it was not practised in Rome. Instead he dissected animals such as pigs and monkeys, which led him into various anatomical errors when drawing analogies with the human body. He developed a threefold theory of physiology which was clearly indebted to Erasistratus: that the brain was the seat of the rational faculties, and controlled the nervous system through the psychic *pneuma;* that the heart was the seat of the passions and conveyed life-giving arterial blood and *pneuma* through the body; and that the appetites were centred in the liver, the sources of the venous system which nourishes the body. This understanding of the body's functioning became enshrined in Christian and Islamic science for centuries. Galen found evidence of purpose and design everywhere in the human body, and endorsed the Platonic doctrine of the *demiourgos* – the craftsman – to such an extent that his medical doctrines tended towards a natural theology, and this in part explains his enormous attraction for later physicians working within a religious framework.

CLASSICAL TWILIGHT

After AD 300, knowledge of the precise sciences was preserved in certain academies, above all in Alexandria, where scholars such as Pappus, Theon and Hypatia fulfilled a vital role in transmitting the works of Archimedes, Euclid and Ptolemy, and where a new form of Platonism was born in the third century. Yet there is no doubt that natural philosophy in the west now entered a period of stagnation: how many significant individuals and significant discoveries stand out during the years AD 400–1100? In Alexandria itself, deep tensions grew up between the pagan, Christian and Jewish communities, and mob violence against cultural centres occured periodically; the murder in AD 415 by an enraged mob of Hypatia, the first woman in the history of science, seems a fitting symbol of the end of the philosophic age.

What were the identifiable causes of this historic ebbing of scientific and philosophical thought? First, and most obviously, there was the political instability of Rome itself: threatened with barbarian invasion and rent by discordant factions, intellectual life could hardly be expected to flourish, while the growing estrangement of the eastern and western sections of the empire cut Rome off from the centres of Greek learning – Alexandria, Athens and Byzantium (for example the legacy of Ptolemy in astronomy, geography and optics was never translated into Latin for a Roman audience). The greatest event in the intellectual history of this period was the emergence of Christianity as the majority religion. Theoretically, the doctrine of an omnipotent creative deity might be expected to encourage the study of form and design in the created order, but that kind of rational theology lay in the future. The thinkers of the early church were more occupied with the problems of the trinity, of human sin and free will, of the authority of the church, and other doctrinal matters, than with researching into mathematics, human

Dioscorides and an artist work on the great medical herbal *Materia medica*, watched by Epinoia, the spirit of inventiveness.
Vatican Library, Rome

An imaginary portrait of Ptolemy studying his astrolabe, from a fifteenth-century manuscript. The crown represents a persistent confusion with the Ptolemaic kings of Egypt.
Biblioteca Marciana, Venice

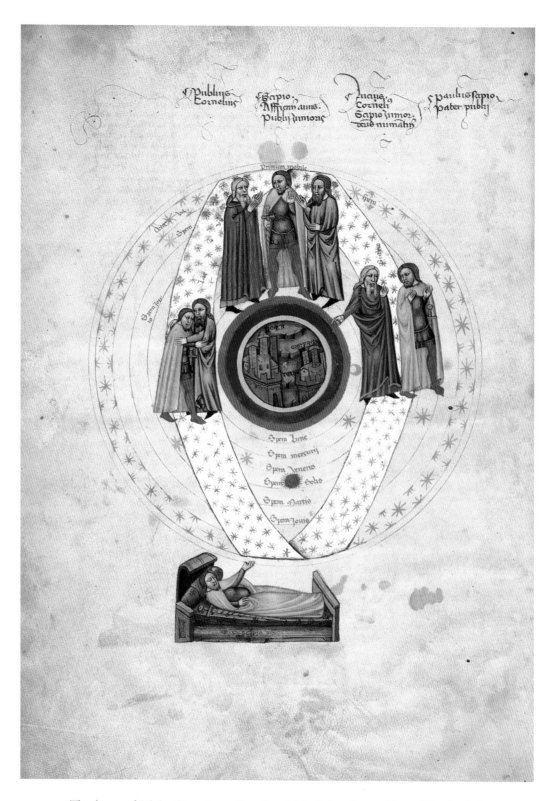

The dream of Scipio. Cicero's text *Somnium Scipionis* describes a vision of the classical
cosmic spheres, inhabited by the souls of the dead.
Bodleian Library, Oxford

physiology or astronomy, especially when these subjects could be regarded as part of the tainted legacy of paganism. And even had the will existed, there was simply no means to advance the study of nature, no instruments to measure and analyse the composition of matter or the movements of the heavens. So the ebbing of innovative natural philosophy in the late Roman era was, if not inevitable, at least understandable. A condensed version of classical science was preserved in the west in a handful of key Latin texts. Macrobius (early fifth century AD) wrote a commentary on Cicero's *Dream of Scipio*, in which he outlines the structure of the celestial spheres, Pythagorean number theory and Ptolemaic geography, including the sphericity of the earth. Martianus Capella (*c*.AD 365–440) was the author of an imaginative allegory entitled *On the Marriage of Philology and Mercury* in which seven bridesmaids each offer descriptions of their intellectual disciplines, and in the process some of Euclid's geometry, Pliny's geography and Ptolemy's astronomy is presented. These two works exercised an enormous influence on the cosmology of the medieval west, providing for example the structure behind the journey through the celestial spheres in Dante's *Paradiso*. The pre-eminent scholar of this twilight period was Boethius (*c*.AD 480–525), whose avowed but unfulfilled aim was to translate into Latin the entire works of Plato and Aristotle. Boethius laid no claim to scientific originality, but he was responsible for the transmission of important elements of Greek logic and a simplified version of Euclid's mathematics. His most personal work, *De Consolatione Philosophiae*, composed while under sentence of death, was said to be, after the Bible, the most widely-read book of the entire middle ages, popularising a form of Neoplatonism which was seen as complementing the Christian religion. While avoiding any specific reference to Christian doctrine, this famous work taught that the intellect can achieve certainty about the existence of God and his goodness, and that the universe is an orderly chain of cause and effect, even though the connecting links in that chain may be invisible to the human mind.

If Hellenic science had been reduced to brief summaries and commentaries, was the legacy of Greek thought of no lasting value, was it a glorious interlude, after which modern science was compelled to re-invent itself? On the contrary, there can be no doubt that the terms in which Greek thinkers approached nature has been of permanent influence. They defined fundamental problems of the nature of matter, the structure of the cosmos, the nature of life, and set out to study them in rational, secular terms, through observation, measurement and deduction. Inevitably these studies led on to questions of cause and design, which in turn signalled a return to theistic answers. In Plato, Aristotle, Ptolemy, Galen and many others, the question of design is either explicit or implicit in all their theories: if order is recognised, does that order imply design? And is that design inherent in matter, or imposed from outside? The atomism of Democritus and Lucretius was a minority voice, an avowed rejection of all gods and causes, and although long eclipsed and neglected, it too was prophetic of future ideas. The recognition that nature operates through ordered systems, the discovery of a mathematical language in which to describe them, and the argument between randomness and design – these are the permanent legacies of the classical achievement.

Chapter Three

SCIENCE IN RELIGIOUS CULTURES

PART 1. THE ISLAMIC MASTERS

&❈

'Hundreds of views exist about the world,
And God, who is ancient, is only the latest'.

Abu al-Ma'arri, 11th century AD

THE ABBASID COURT

It is related that in a dream the great Abbasid Caliph, al-Mamun, saw a vision of a blue-eyed, noble-headed man, reclining on a couch. To the Caliph's question 'Who are you?' the figure replied 'Aristotle'. The Caliph was delighted, and proceeded to discuss goodness, law and faith with the great philosopher. This legend perhaps surprises us, and we wonder why a ruler of the fiercely religious world of Islam should find inspiration in an ancient pagan philosopher. In fact the legend embodies one of the central intellectual events of the middle ages: the process by which the achievements of Greece became part of Islamic culture, and were in turn transmitted to the west.

What happened to Greek science as the Roman Empire decayed? In AD 549 the Emperor Justinian closed the pagan schools in Athens, an event that is taken as symbolizing the end of the classical era; but the links between Roman and Greek thought had been severed much earlier, and knowledge of the Greek language became virtually extinct. The blueprint of basic education in western Christendom was provided by Martianus Capella: the three arts of grammar, rhetoric and logic dealt with literary study, while four further disciplines – arithmetic, geometry, music and astronomy – embodied all of the science that could be formally studied. Basic elements of the science of Euclid and Ptolemy would be further condensed by Christian encyclopedists such as Cassiodorus and Isidore, while the study of theology blossomed and absorbed virtually all the intellectual energy of the west. This decline of science was matched by the decline of urban society, the shift towards a predominantly rural culture, with learning centred in the monasteries. In this post-classical world, learning was overwhelmingly religious and literary, not scientific. In the east however, in Athens, Alexandria and Antioch, where the language barrier did not exist, the texts of Aristotle, Plato, Euclid, Ptolemy, Hippocrates and Galen were known and read, even if they no longer inspired a living tradition of original thought. At the eastern margins of the empire too, intellectual life flourished: in the cities of Sassanian Persia – Ctesiphon, Edessa, Harran and Seleucia – Christianity mingled with Zoroastrianism, Neoplatonism and Vedic thought from India. Certain elements of Greek science are known to have spread through this region, and Ptolemaic astronomy for

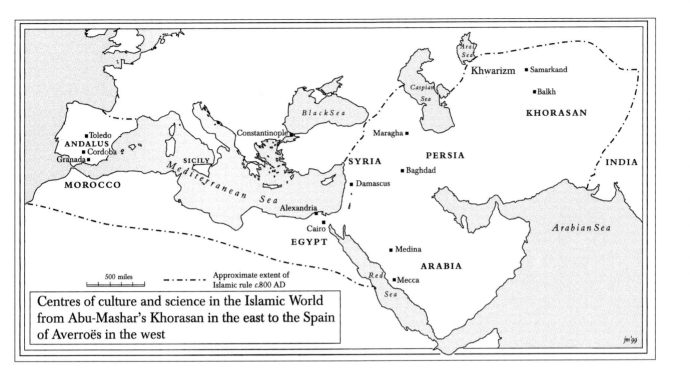

Centres of culture and science in the Islamic World from Abu-Mashar's Khorasan in the east to the Spain of Averroës in the west

Centres of culture and science in the Islamic world. Islamic science drew on many traditions – Greek, Sanskrit and Persian – and Arabic functioned as a *lingua franca* to unite this vast empire.

Map by John Mitchell

example had reached India by the fifth century, where the geometric movements of the planets were used to plot astrological positions.

Into this intellectual mélange swept the ferocious tide of Arab conquest, so irresistible that by AD 750 all the cultural centres of the east, except Constantinople itself, were in Islamic hands. It would hardly have been surprising if the all-powerful religious conquerors had deliberately destroyed the pagan elements of the civilisations which they now controlled, but they did not; instead they set out to absorb what they recognized as being, in some important way, higher cultures. This process is seen at its clearest at the Abbasid court in Baghdad, where the Caliph al-Mamun (r. AD 813–833) established the 'House of Wisdom', dedicated to the translation into Arabic of Greek philosophical and scientific works. It has been said that one the chief scientific inventions of Islam was patronage, for nothing like this form of political direction in intellectual life had ever been seen in Greece or Rome. Greek texts were gathered from Constantinople, Antioch and Alexandria and taken to Baghdad in a highly international enterprise, with many of its leading figures being Christians, Jews or pagans, and in addition, works were translated from Persian and Sanskrit. Syriac, the cultivated tongue of the Middle East at this period, was often an intermediate stage between Greek and Arabic. The intellectual context of this whole extraordinary process was the movement known as *Mu'tazilah,* whose adherents held that unaided human reason was capable of discovering truth and goodness, and that the revelations of the Koran were not the sole pathway to divine truth. This form of rational faith became virtually the official religion of the caliphate in the ninth century, but met periodically with strong resistance and reaction.

The most famous translator was Hunayn Ibn Ishaq (died AD 873), a Christian who served at the House of Wisdom, translating Plato, Aristotle, Hippocrates and Galen into Syriac, which was then rendered into Arabic. Hunayn's son Ishaq (died AD 911) made all-important Arabic versions of Euclid's *Elements* and Ptolemy's *Almagest*, which were immediately recognized as the supreme authorities in Islamic mathematics and astronomy. It was the process of translating these fundamental works that raised Arabic to the status of the *lingua franca* of a new science. The versatility of the scientists in the House of Wisdom was truly astonishing, and this remained a characteristic of Islamic scholars. A figure such as Thabit Ibn Qurra (AD 836–901) mastered the exact sciences of mathematics and astronomy, was a practising physician and prolific medical author, and still made serious contributions to philosophy, logic and philology.

It is easy to see why Islamic rulers, like all rulers in the ancient and medieval world, should be interested in cultivating sciences such as astrology and medicine. Moreover certain practices within Islam itself served to encourage an appreciation of precise science. One was the need to devise a new calendar dating from the *Hegira* of Muhammad in AD 622, and to regulate the prescribed times of prayer, while the other was the *Qibla,* the sacred direction of Mecca, which must be ascertained for the purposes of prayer; both these objectives could only be achieved through a mastery of astronomy, and both became the subject of an extensive literature. In one wall of every mosque, a niche called the *mihrab* marked the direction in which worshippers faced during prayer. Prayer times were determined by solar altitude and by the length of shadows, and this information was embodied in written tables, while the *Qibla* involved a knowledge of coordinate geometry and trigonometry. These were important technical matters and they gave mathematics and astronomy a high status in Islamic culture. However, it must be asked whether they, or the rational faith sponsored by the caliphate, led further, to a form of scientific thought which was in some way original or distinctive: when we speak of Islamic science, was it in fact specifically Islamic, or is that simply a convenient historical label? Nowhere is that question more pertinent than in the case of the astrologer Abu'Mashar (AD 787–886), one of the earliest and most celebrated of all Islamic scientist-philosophers, and one who profoundly influenced the west through his Latin persona 'Albumasar'.

THE IMPACT OF ASTROLOGY

In the history of science astrology is of absorbing interest as a body of knowledge once considered genuinely scientific – by no less a figure than Ptolemy for example – but later outmoded and discredited. Some form of divination from the stars was found in almost all ancient cultures, and derived initially from the conviction that the heavenly bodies were intelligences or deities. These deities possessed the power to influence the world of man, to cause war, death or good fortune, and their will was believed to be discernible from their elevation in the heavens. The notion that planets might control human destiny *as planets,* by virtue of their astronomical positions alone, is a later belief. It raised the difficult problem by what process their power reached the earth and operated. This was a physical question, and a physical answer was available in the cosmology of Aristotle. In the enclosed system of planetary spheres, just as the motion of each sphere was imparted by its neighbour, so a less visible but no less powerful force, able to shape human destiny, was transmitted through the layered spheres. Ultimately, the precise nature of this force eluded all definition, but the enclosed cosmic spheres of Greek science did provide a convincing field in which it might operate, while the circular motion of the concentric spheres undeniably brought about the all-important periodic conjunctions of the heavenly bodies. Using this model of the cosmos, it was the fundamental task of the astrologer to chart these conjunctions

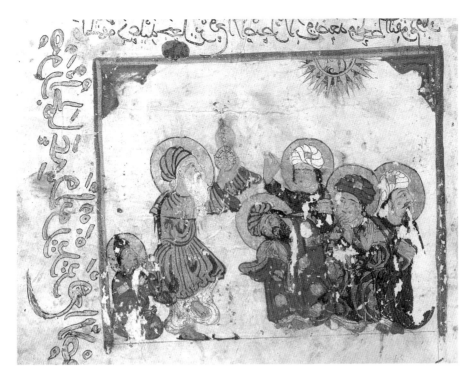

An Islamic astrologer instructs a group of pupils in the use of the astrolabe, from a thirteenth-century manuscript.
Bibliothèque Nationale, Paris

in the past and the future, and to explain what their effects would be, whether on individuals or on nations. The first part of this task was genuinely scientific, requiring advanced geometry to calculate the required positions. By contrast the second part was based on a semi-religious, semi-legendary corpus of beliefs about the celestial deities and their ability to engender love, war, wisdom or riches.

It is plainly a considerable problem to understand how such a system of *animistic* beliefs could co-exist with a revealed, prophetic religion, whether Islam or Christianity. The early fathers of the Christian church did indeed condemn astrology as a variety of paganism. Yet it never entirely disappeared, and in the Christian middle ages serious interest in astrology sprang up again, and stronger than ever before. Curiously, the astrological renaissance in Europe came principally from Islamic sources, and it was the legacy of Abu'Mashar which played a major part. Although he entered the House of Wisdom in Baghdad before he was thirty years old, the place of Abu'Mashar's birth, Balkh in modern Afghansitan, was of great importance. It was a central Asian outpost first of Hellenic, then of Persian culture. Jews and Christians, Buddhists and Hindus, Manichees and Zoroastrians, all were found in Balkh. The city and the region fell to the Arab conquerors in AD 651, but the older intellectual traditions survived alongside the new faith, and thinkers such as Abu'Mashar introduced some of this eclectic legacy to the Islamic world through the *lingua franca* of Arabic. In Sassanian Persia, and still more in the India of the Gupta period, astronomers had absorbed the lessons of Greek science, and had built sophisticated mathematical procedures for plotting celestial movements. In both cultures these procedures were used to underpin an elaborate astrological system. In the Arab heartlands of Islam, a practical form of astronomy was well established with which to measure time and seasons, but it was far below the mathematical astronomy of India or Persia. Abu'Mashar was struck by the richness of these diverse traditions, and developed the doctrine that all intellectual traditions are revelations of a single truth, and he saw this exemplified above all in astrology, with its distinct but closely related systems found in Indian, Persian and Greek culture. In his principal work, the 'Great Introduction to the Science of Astrology', he places the framework of astrology firmly within Aristotelian cosmology. The universe consists of three realms: the divine, which is the sphere of light, of the prime mover beyond the stars; the ethereal, which contains the eight celestial spheres; and the sub-lunar or earthly sphere. Man's soul, according to Abu'Mashar, originally descended from the sphere of light to the earth, and his true destiny is to strive to return to the divine. But man cannot pass through the ethereal realm without the help of the intermediary presences in the celestial spheres. Therefore his intellectual efforts must be devoted to understanding the functioning of these spheres and the qualities of their presiding spirits. In practice, astrological science went on to study a supposed three-way correlation between celestial forces, human attributes and terrestrial things such as plants, animals and minerals. Apparently mystical or occult as this approach appears, Abu'Mashar devoted enormous labour to calculating a new improved *zij* – a manual of mathematics and astronomical data which enabled planetary positions to be calculated – to provide an accurate basis for this study.

In a further work, the 'Book of Conjunctions', Abu'Mashar attempted to apply the science to history, relating the conjunctions of Sun, Moon and planets to the rise and fall of human powers and institutions. He daringly suggested that the Abbasid caliphate, and even Islam itself, might in time be subject to historical decay. Perhaps the most sensational of these historical theories was that the world had been created at the great conjunction of the seven planets in the first degree of Aries, and that it would end when they reached conjunction again in the last degree of Pisces. This concept he clearly derived from his knowledge of the India *yuga,* the immense periods of time over which the heavens were believed to return to their original positions. The translation of Abu'Mashar's representative works into Latin in the twelfth century was responsible for the resurgent western interest in astrology, and his reputation became immense, lasting into the seventeenth century (with the decline in astrology, he was portrayed in a play by Dryden, *Albumazar,* as a knavish Faust-figure).

Where, in this intoxicating intellectual game, did the Islamic religion figure? The answer is that it was to be seen as only one, prophetic, manifestation of the divine. For Abu'Mashar there were other pathways to a knowledge of the divine which were susceptible to scientific investigation, while the purely religious was not. His thought was known and studied throughout the Islamic world, and yet its status as 'Islamic science' is highly questionable. Why then did his work acquire such authority? Behind that question lies the wider one of the continued credence placed in astrology by pagans, Muslims and Christians alike, in spite of its manifest failure to fulfil its promises. This can only be explained by understanding that astrology did indeed make sense of certain problems in the natural and human world. In the first, it interpreted the complex movements of the heavenly bodies by supposing that they expressed a will and an intelligence on another plane from our own; while in the second, the ebb and flow of human fortune, the wealth and power that fell to some and eluded others, might be seen as subject to hidden laws, and the prospect of understanding those laws was indeed an enticing one. Religious men might argue that astrology was not incompatible with divine power, for it merely sought to elucidate laws which God had built into the universe. Ultimately of course, neither the astronomical nor the human mysteries were *explained,* but they were *restated* in terms convincing to that era. Abu'Mashar cannot perhaps be called a great scientist, but he has an important place in the history of science because of the way he synthesized the diverse intellectual forces current at the opening of the Islamic era. Astrology represented science in its exciting, magical, empowering aspect; not Greek, not precise, but oriental, mystical and enticing.

THE ABBASID GOLDEN AGE

Contemporary with Hunayn Ibn Ishaq and Abu'Mashar at the House of Wisdom were the mathematician al-Khwarizmi, the philosopher al-Kindi and the astronomer al-Farghani, whose works constituted the first golden age of Islamic science. The fame of al-Khwarizmi (*c.*AD 790–850) rests on his popularization of two great mathematical principles: the use of algebra and the use of Indian numerals. The very word algebra comes from the title of one of his works 'The Book of *al-Jabr* and *al-Muqabala*', meaning respectively 'restoration' and 'balancing'. The two essential operations of algebra were the elimination of negative quantities, and the reduction of positive quantities on both sides of an equation. Al-Khwarizmi explains the purpose of his work as setting out 'what is easiest and most useful in arithmetic, such as men constantly require in cases of inheritance, legacies, partition, lawsuits and trade ... or where the measuring of lands, the digging of canals, geometrical computations and other objects are concerned.' He explains how to reduce all problems of this kind to one of a handful of standard forms by the processes of 'restoration and balancing'. Thus the form:

$$50 + x^2 = 29 + 10x \quad \text{is reduced to} \quad 21 + x^2 = 10x$$

Like the Egyptians and Babylonians before him, al-Khwarizmi had not arrived at anything approaching modern algebraic notation, but expressed his calculations in verbal, rhetorical form, in steps which we would call an algorithm – a word which is a corruption of his own name. Thus the problem which we would write as:

$$(^x/_3 + 1)(^x/_4 + 1) = 20$$

al-Khwarizmi expresses as 'A quantity: I multiplied a third of it and a *dirham* by a fourth of it and a *dirham*: it becomes twenty' (the *dirham* was a unit a coinage, here representing 1). Further algorithms are given for finding plane areas and solid volumes, and although the calculations are not on a very profound level, the elaboration of these formulae stood at the beginning of a new devlopment in mathematics. This is perhaps even more true of al-Khwarizmi's other innovation, the introduction of Indian numerals, which we call Arabic numerals, into Arabic science. The Arabs had previously used an alphabetic system, in which numerals were represented by different letters, and higher values were made up of compounds which proved unwieldy for calculation, like the Roman system. By the sixth century AD the decimal place-value system was in use in India, and al-Khwarizimi was the first to expound it systematically to his Islamic contemporaries. It is usually said that al-Khwarizmi introduced the use of the zero into mathematic, but Greek mathematicians, and others, had marked the absence of a finite number in various ways, and what al-Khwarizmi did was to place the zero firmly in the context of the Hindu-Arabic place-value system. A century after al-Khwarizmi, the Damascus mathematician, al-Uqlidisi, employed the Indian numerals in a seminal treament of decimal fractions. It is easy to see why the flexibility of the Indian numerals and the decimal system led eventually to the discarding of the abacus in favour of paper and ink calculation. Al-Uqlidisi's

The Pythagorean theorem in a manuscript of a work by al-Tusi.

The British Library
Add.Ms. 23387 f.28r.

name indicates that he was a professional scribe who copied and sold manuscripts of Euclid's works – an intriguing demonstration of the importance of mathematics in Islamic society.

Al-Kindi (*c*.801–*c*.866) is traditionally known as the 'first Arab philosopher', and he directed enormous effort to reconciling rational philosophy with revealed religion. He taught that the truth of the Greek philosophers was the same as that of the prophets, but that the latter was given at sudden moments by the divine will, while the former was discovered gradually by the disciplined exercise of reason. The purification of human life is the path offered by both philosophers and prophets, and the ultimate goal was the apprehension of the divine. Al-Kindi was a prolific researcher, and wrote extensively on optics, medicine, pharmacology and musical theory, seeking always to preserve the discoveries of the past, and test them in the light of reason. If al-Kindi can be regarded as a founder of Arab philosophy, then al-Farghani (*fl*.AD 830–860) occupies a similar position in astronomy. He produced a summary of Ptolemaic science so clearly and attractively presented that it was still being used in the west in printed Latin editions in the sixteenth century. In non-mathematical language al-Farghani dealt not only with Ptolemy's planetary system and star catalogue, but with his theory on the size of the universe, and with his terrestrial geography. He also composed one of the most extensive descriptions of the astrolabe, the instrument which was widely used by the astronomers of the ninth century, and whose elegance and precision has made it a fitting symbol for the science and ingenuity of its Islamic masters.

The astrolabe was a hand-held instrument, between four and twelve inches in

diameter. It consisted of two flat metal plates designed to be superimposed, one on top of the other. The top plate, called the *rete*, meaning 'net', was fretted into an open framework bearing a number of precisely placed pointers, which marked the positions of certain bright stars. On the lower plate were incised lines representing local celestial cordinates, that is, altitudes and azimuths in relation to the horizon at a particular latitude. When the rete was placed over the lower plate, the stars were effectively located in the sky, and the instrument became a map of the northern heavens. The rotation of the rete about the centre point, representing the celestial pole, brought into view the stars which would be visible at any given time or season. The astrolabe plate, with its lines of celestial altitude and azimuth, was valid for one terrestrial latitude only, but it could be replaced by others constructed for different latitudes. A type of universal astrolabe, valid for all latitudes, was later developed, but it was more complex to use, and was far less common. The principle on which the astrolabe worked was the stereographic projection, in which the stellar positions are projected onto the plane of the celestial equator, just as a map of the spherical earth may be projected onto a plane surface. The theory of this projection was well known to Ptolemy and other Greek astronomers, and it seems likely that instruments resembling astrolabes were actually made in the classical period, although no examples have survived. Islamic astronomers used the instrument to work out the positions of Sun, Moon and stars relative to the horizon and the observer's zenith at any given time. It had important astrological functions, for it gave the degrees of the ecliptic on the eastern and western horizons and crossing the meridian – the ascendent, the

Astrolabe made in Egypt or Syria in 1235 AD. The star pointers have been shaped into decorative figures and foliage. The lines of celestial altitude are clearly visible on the lower plate, and this instrument was made with plates for the latitudes of Cairo and Baghdad.
British Museum, Dept. of Oriental Antiquities.

descendent, and the mid-heaven – which were essential in casting horoscopes. The astrolabe was in fact an analogue or working model of the heavens, and its construction and use sprang from a very precise understanding of celestial movements and of spherical geometry. The other great tool of the astronomer was the table of mathematical formulae for calculating celestial positions known as the *zij*. The mathematical basis of the *zij* was derived from Ptolemy, and it spread both to India and throughout the Islamic world. It would include celestial geometry, solar and lunar tables, planetary motions, a star catalogue, some terrestrial geography, and the fundamentals of astrology. New *zijes* were compiled from time to time by leading astronomers, using improved observations or methods of calculation. Taken together, the astrolabe and the *zij* demonstrate the enormous superiority of Islamic over western astronomy between the eighth and the thirteenth centuries. It was from the astrolabe treatises and the *zijes,* when they became known in Europe, that the Arabic astronomical vocabulary – zenith, azimuth, nadir etc. – entered western science.

PERSIANS, ANDALUSIANS AND OTHERS

Not surprisingly, the importation of pagan science into Islamic civilization, and especially the central, philosophical importance accorded to it by figures such as Abu'Mashar and al-Kindi, provoked opposition among orthodox believers. In the eleventh century AD a systematic refutation of the methods of the *falasifa* (there was no equivalent Arabic term for 'philosopher') was launched by the mystic al-Ghazali in his *Destruction of the Philosophers*. Still more hostile was the jurist

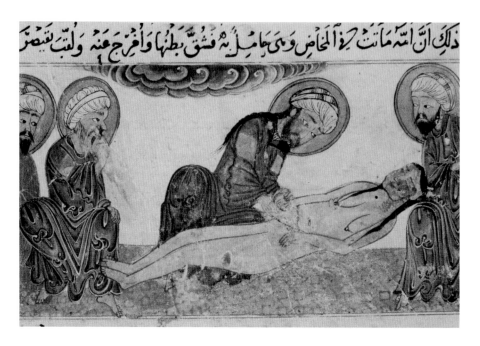

An operation, possibly a Caesarian section, being performed on a woman. From a manuscript of a work by al-Biruni.
Edinburgh University Library

Ibn Taymiyya in the early fourteenth century, who argued that Greek principles of science and logic were wholly irreconcilable with revealed religion. Yet the first scholarly experiment in Baghdad was no brief golden age, and the party of reason out-manoeuvred the party of orthodoxy to the extent that Islamic science flourished for five hundred years, spreading to other centres, to Cairo, Persia, Central

Asia, Sicily and Spain. In Persia the first great Islamic physician, al-Razi (*c.*AD 854–925) who was known to the west as Rhazes, directed hospitals, and wrote comprehensive medical treatises in which he compared descriptions of diseases in classical authors with his own observations. He was a follower of Galen, but a critical one, indeed he entitled one of his works *Doubts Concerning Galen*. Al-Razi taught an anti-authoritarian approach to science and to philosophy, arguing that even the greatest of past masters had achieved only provisional results. For example, he disagreed sharply with Aristotle's axiom that the universe was finite, for reason teaches us that even if the bodies which define space are absent, then absolute space still exists. Al-Razi was disdainful of authoritarian religion, holding it to be an affront to man's intellectual freedom; that such a heretic could write and teach unmolested says much for the tolerance of Islamic intellectual life. Like Abu'Mashar, we feel that al-Razi is part of Islamic science through the time in which he lived and through the language in which he wrote, but not through his culture or his convictions.

Al-Razi's more famous Persian successor Ibn Sina ('Avicenna' to the west) also combined medicine with philosophy, but a philosophy of a deeply religious kind. Ibn Sina (AD 980–1037) sought to reconcile the religion of the Koran with both an Aristotelian attention to nature and a Platonic idealism. Religion and science

Islamic medicine was superior to that of Europe throughout the middle ages and Renaissance, but its anatomical illustration remained imprecise and stylized; this kind of figure showing very generalized blood, digestive or nervous systems, was drawn to illustrate Islamic medical works down to the eighteenth century.
Bodleian Library, Oxford

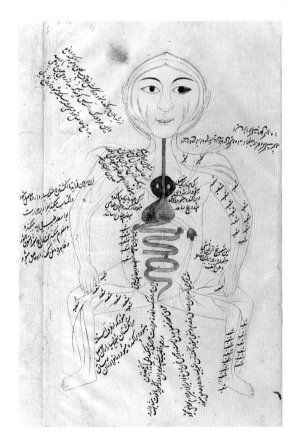

were twin pathways to the divine, for all things emanate from God and desire to return to God. He devised a complex metaphysical system linking the physical properties of the universe – time, movement, generation – with each other and with their divine creator. The entire tendency of his thought is to analyse into metaphysical categories the things of this world and the supposed forces which act upon them, and it is easy to see why orthodox believers were deeply suspicious of this highly intellectual approach. As a physician Avicenna was outstanding for his massive attention to detail rather than for any original discoveries. He accepted the Galenic doctrine of the four humours, and the three bodily systems. His great book *al-Qanun* (the 'Canon' or code of medical laws) extended to one million words, treating hygiene, therapy, drugs and surgery, and enumerating the symptoms of disease of every part of the body, literally from the head to the toes. It was translated into Latin in the twelfth century, printed first in 1473, and remained a standard medical textbook in some European universities until the seventeenth century.

Destined for even greater permanence was the work of the Persian astronomer al-Sufi (AD 903–986), who fixed the names of several hundred stars, many of them still in use. The Greeks had given names to only a handful of stars, such as Sirius and Arcturus, while the Arabs had traditional names for many more, characteristically those beginning with the prefix *al,* such as Aldebaran and Algol. Al-Sufi's work of naming the stars occurred in the context of his revision of Ptolemy's star catalogue from the *Almagest,* and many of the names are derived simply from the star's place in the constellation; for example Mintaka in Orion means 'belt', Markab in Pegasus means 'shoulder'. Al-Sufi revised Ptolemy's data on star positions, magnitude and colour, and corrected all the celestial longitudes to allow for precessional movement to the epoch AD 964. Al-Sufi's *Book on the Constellations of the Fixed Stars* was frequently re-copied, adorned with elegant pictures of the constellations, and its author became well known in the west as Azophi. The last of these great Persian scientists famous in the west was Umar al-Khayyami (AD 1048–1131), known to us as the poet Omar Khayyam, who supervised the observatory at Isfahan for many years, and effected important calendar reforms. He was an original mathematician who expounded and clarified many of Euclid's methods, made important advances in geometric algebra, and contributed to the theory of musical intervals. His poetry is an extraordinary blend of fatalism, cynicism and sensuous aestheticism that defies classification, and certainly brought him in his own day under suspicion of atheism.

The Persian genius for subtlety and precision of thought took a practical form in the work of al-Khazini (*fl.*AD 1110–1130) who made and described scientific instruments, most notably a superb hydrostatic balance which was used to weigh gold and gems. It was called by its maker the 'Balance of Wisdom' and had a claimed sensitivity of one part in 100,000 for a weight of around four kilograms. Al-Khazini was something of a mystic, and wrote of his balance as a symbol of the divine justice which distinguishes the true from the false, the precious from the worthless. He was able to measure the specific gravities of many metals and other substances with great accuracy, and his work with the balance led him into speculations about physical forces, which, although inconclusive, were advanced for their time. Less elevated but no less intriguing was the career of al-Jazari (*fl.*AD 1200), designer of machines and automata in the tradition of Hero of Alexandria. Water-clocks, trick drinking vessels, fountains, locks and pumps are all displayed in al-Jazari's 'Book of Knowledge of Mechanical Devices'. As in the case of Hero, the designs are ingenious, but the principles were never extended into wider use. Was it that the limiting factor lay in the understanding of power? Sufficient power could be channelled to achieve a small-scale effect, but if that were magnified beyond a certain point, the apparatus would fail. Or was it simply that there was no concept of the machine as a working tool, no tasks for which muscle would not suffice?

The influence of Greek models in Islamic science was paramount, but there were dissenting voices, and even the towering figure of Ptolemy found a stern critic in Ibn al-Haytham (AD 965–1040) who argued that certain of Ptolemy's geometric models – the equant and the eccentric – violated the principle of uniform circular motion, and were impossible to reconcile with physical reality.

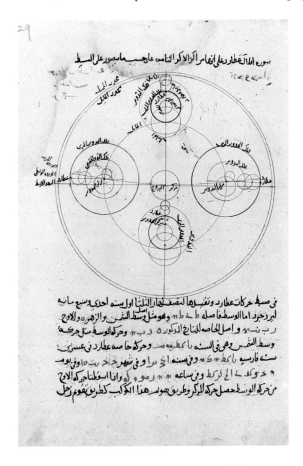

Al-Haytham was at his most original in his theory of light and vision, where he proceeded by experiment and induction, and again voiced his scepticism of past authorities. He rejected the classical concept of the visual cone emanating from the eye, pointing to the absurdity of supposing that some form of radiance could flow from the eye the moment it was opened and illuminate the entire heavens. Instead he affirms clearly that primary light from luminous bodies is received by all other objects, which in turn become luminous; al-Haytham thus took the crucial step of ascribing the visibility of objects to reflected light. Al-Haytham was the first to employ the concept of rays of light, travelling in straight lines from every point on its source in all directions; this was a vital advance on the classical theory that objects emitted coherent images of themselves. Colours he considered distinct from light, but always associated with it, and behaving in precisely the same way as light. Light rays are focused by the lens, and forms of objects take shape inside the eye, to be transmitted to the brain via the optic nerve. He gave mathematical analyses of reflection, refraction and the geometry of the *camera obscura*. His work remained definitive for centuries in the east and the west, where Latin editions of 'Alhazen' were studied by Kepler, Fermat and Descartes. Three centuries later, al-Haytham's work was extended by Kamal al-Din (d. AD 1320), who succeeded where others had failed in explaining some of the basic geometry

The path of the planet Mercury: a complex hypothetical model of the Ptolemaic type, employing multiple epicycles, from a work of the fourteenth century by Ibn ash-Shatir.

Bodleian Library, Oxford

of light in the rainbow. By experimenting with glass spheres he showed that sunlight was twice refracted and once reflected inside raindrops. He was not able however to analyse the colours as constituents of white.

The great channel through which Islamic science, and its Greek models, was to reach the west was Spain, the Muslim kingdom of al-Andalus. Here the Emir al-Hakam II (*r.*AD 961–976) rivalled the achievements of Baghdad a century earlier in patronising the arts and sciences. Yet al-Andalus was also the scene of a constant battle between religion and rationalism, and after al-Hakam's death a reaction set in, during which scientific books were burned and philosophy was driven underground. With a change of dynasty, a revival began, and most of the great scientific figures of al-Andalus belong to the twelfth century. Intellectually, Andalusian science became strongly Aristotelian, culminating in the massive commentaries of Ibn Rushd, or Averroës (AD 1126–1198). While personally devout, and accepting the revealed religion of the Koran, Averroës also believed that the world of nature and the structure of human thought could only be understood in terms of Aristotelian logic, metaphysics and science. Much exercised by the Aristotelian belief that the world was eternal, Averroës proposed to reconcile this with religious doctrine through the concept of continuous creation, a process which allowed for absolute continuity and was always dependent on God, but within which new forms in nature could constantly emerge. Only God himself was self-sufficient and without cause. This ingenious type of evolutionism was of great interest to later western thinkers. Both Averroës and the great Cordoban astronomer al-Bitruji (*fl.*AD 1190) were deeply dissatisfied with the complexities of Ptolemy's model of the universe as presented in the *Almagest*, because it violated Aristotelian physics. Averroës wrote 'The astronomical science of our day surely offers nothing from which one can derive an existing reality. The model that has been developed in the times in which we live accords with the computations, not with existence.' Rejecting the Ptolemaic devices of epicycles and eccentrics, al-Bitruji put forward an alternative model of spiral celestial movements which were perfectly circular, with each sphere tangent to the next. Western thinkers such as Albertus Magnus were greatly interested in this alternative, and debated its merits against the Ptolemaic model.

Contemporary with Averroës and al-Bitruji was the great geographer al-Idrisi (AD 1100–1166) who was educated in Cordoba but whose most important work was done in Sicily, where the Norman king, Roger II, presided over a distinctly international court. It was at Roger's command that al-Idrisi compiled the most detailed world map of the middle ages, based largely on Ptolemy, but supplemented with data from the vast Islamic realms and from Europe. The known world – Europe, most of Asia, northern Africa – was divided into seven latitudinal climates, and into ten longitudinal sections, and for each of these seventy regions a detailed map was compiled, with accompanying texts describing lands and people. The original map is thought to have been engraved on silver, but manuscript copies were soon made. This scale of world map was far in advance of anything known in the west at this period. The work of al-Idrisi is of special interest in the contrast it presents with the traditional Islamic geography, where only the Islamic empire was mapped, and whose world maps were spare and diagrammatic, and centred on Mecca rather as the Christian *mappae mundi* were centred on Jerusalem. The more secular approach to geography had been advanced in the work of the Baghdad scholar al-Masudi (d.AD 956), who travelled the entire known world, and wrote historical and geographical works in which the legacy of Greece, Rome, India and Persia were given a weight equal to that of the Islamic nations.

It was earlier, before the religious conflict of the eleventh century, that the great medical authority in Muslim Spain, Abu'l Qasim (*c.*AD 936–1013), flourished, the Abulcasis of the west. He composed a massive medical encyclopedia, of which the surgical part was the most advanced in the middle ages and the most admired. He designed many new instruments and has left the first accurate descriptions of the removal of aural polyps, of female lithotomy and hydrocephaly. He was also an expert druggist and discussed the preparation of mineral and vegetable medicines. He recognized very clearly the psychological dimension in healing, and occasionally advocated the use of mood-altering drugs such as opium.

It was inevitable that the various regions of the Islamic realm, which stretched from Spain to Afghanistan, would diverge intellectually, and the patronage of the regional court was all-important. While Spain developed its Aristotelian edifice, further east newer currents, of astronomy especially, had emerged. A tradition of precise observation was established by al-Biruni (973–*c.*AD 1050), a native of the ancient region of Khwarazam south of the Aral Sea. Known to contemporaries as 'the Master', al-Biruni was not the author of new theories or discoveries, but his range of research and writing was enormous, especially in astronomy and geography, chronology and mathematics. He also mastered Sanskrit and wrote a major description of the geography and civilization of India. A devout Muslim himself, his deep interest in the science and history of other cultures is a paradigm of the Islamic passion for rational synthesis.

A century after al-Biruni, another renowned philosopher-scientist, the Persian al-Tusi (AD 1201–1274) was caught up in the Mongol assault on the Islamic world, and was invited, or perhaps compelled, to enter the service of Hulagu, the grandson of Genghis Khan and conqueror of Baghdad. Although known to history as a great destroyer of Persian culture, Hulagu had a great respect for al-Tusi, possibly inspired by his own interest in astrology. A major observatory was built at Maragha in north-western Persia, which al-Tusi supervised from 1259 to 1274, using the large naked-eye sighting instruments to compile new, accurate astronomical tables. Observatories with such equipment were of course unknown in the Latin west at this time. Al-Tusi devised new mathematical models to account for celestial movements which Ptolemy could not convincingly explain. He composed works on mineralogy, giving alchemical accounts of the formation of metals, and wrote on logic, ethics and theology, indeed in the Islamic world he is known as much for his theology as for his science. His passion for knowledge was blended from precise observation, mathematical analysis and an almost mystical quest for the divine.

An observatory of equal historical interest was established at Samarkand in 1424 by Ulugh Beg, grandson of the infamous Tamerlane, but whose attention was fixed on science instead of conquest. The centrepiece of the observatory was a huge arc cut one hundred feet deep into a hillside in the meridian plane, with which the elevations of celestial bodies could be accurately read. With this and with other instruments, Ulugh Beg corrected many existing astronomical values and compiled a new star catalogue. The death of this princely scientist was violent and mysterious: in 1941 his skeleton was discovered by archaeologists, clothed in martyr's robes, and he had clearly been beheaded, but why and by whom has never been established.

The most important Islamic achievements were clearly in the precise sciences, but what of natural philosophy, and the science of the material world? Curiously, the only science which has an Arabic name – alchemy – is the one about whose role in Islamic culture we know least. Alchemy was a form of occult learning less widespread in the ancient world than astrology, and less precise in its aims and

The great mathematician al-Tusi directed the observatory founded
at Maragha (in modern Azerbaijan) by the Khan Hulagu, grandson of Ghenghis Khan.
This fifteenth-century manuscript shows al-Tusi and his assistants at work.

The British Library Or.Ms. 3222 f.105v.

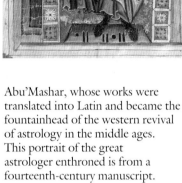

Abu'Mashar, whose works were translated into Latin and became the fountainhead of the western revival of astrology in the middle ages. This portrait of the great astrologer enthroned is from a fourteenth-century manuscript.
Bibliothèque Nationale, Paris

In Abu'Mashar's *Book of Conjunctions* personified figures of the Moon and Jupiter meet in the constellation Sagittarius. Abu'Mashar expounded the central doctrine that these conjunctions focussed astrological forces and acted as turning-points in human affairs.
Bibliothèque Nationale, Paris

The constellation Perseus from al-Sufi's *Book of the Fixed Stars,* from a tenth-century manuscript.
Based closely on the star catalogue in Ptolemy's *Almagest,* Al-Sufi's illustrations located more
than 1,000 stars, and habitually showed the constellations in double aspect:
as seen from the earth, and reversed as on a celestial globe.
The British Library Or. Ms. 5323 f.21v.

methods. Its central inspiration was the recognition that the material world is composed of a myriad of substances which differ from each other, yet are clearly related in certain groups, and which may sometimes combine and change. Alchemy was the attempt to unlock the forces which defined these substances, and then to transmute them. This principle became, or was perhaps from the outset, overlaid with the adept's own ambition to exercise occult powers, to bend nature to his will, and the symbol for these powers came to be the making of gold. The earliest alchemical writings have been attributed to the Egyptian Zosimus, around AD 300, and a large compendium was collected in Byzantium by the seventh century. Many of the processes described for the mingling and transmutation of substances end with the formula 'One nature rejoices in another nature; one nature triumphs over another nature; one nature masters another nature.' Alchemical texts must have passed into Arabic hands, along with scientific and philosophical ones, when cities in Syria, Egypt and Persia were conquered. The subject was explored by leading intellectuals such as al-Razi, who was more interested in its physical than its occult dimension, but the classic alchemical texts in Arabic were attributed to Jabir Ibn Hayyan. Like Hippocrates, Jabir (known to the west as Geber) is an elusive individual, and his name may have been simply a focus for a school of thought. Jabir claimed to have reduced all substances to two – mercury and sulphur – which are the bases of all others. These were however not the chemicals we know, but two principles embodying the four qualities, mercury being cold and wet, sulphur hot and dry. Moreover the materials of the earth had a cosmic significance as talismans of the heavenly bodies: gold was the substance of the Sun, silver of the Moon, copper of Venus, and so on. The process of transmutation into gold was capable of a symbolic interpretation, as the operation of a spiritual power raising human nature into communication with the divine. Whether the practitioners were indifferent to the gold itself is open to question. Jabir was concerned not only with the mineral kingdom but with the living world too, and he entertains the possibility of creating life itself from its constituent elements, thus placing the adept on a level with the divine. Jabir was able to place these ambitions within a religious framework through the allegorical interpretation of certain Koranic passages on the unity of nature, but it is easy to see why such an endeavour would antagonize the orthodox, and the alchemical route to nature's secrets never became central to Islamic science. On a practical level, the Islamic applied arts make it clear that their craftsmen had considerable expertise in chemistry: metallurgy of a high order, the dying of fabrics, and the glazing of ceramics were hard-won skills. When chemical and alchemical texts were translated into Latin, there were no equivalents for many technical terms, hence the Arabic origin of words such as alkali, alcohol, camphor, realgar, talc and borax.

SCIENCE IN A RELIGIOUS CULTURE?

To sum up the nature of Islamic science is exceptionally difficult because certain central questions stubbornly refuse to be answered. How homogeneous was Islamic thought – were scholars in Spain and Egypt, in Arabia and Persia truly aware of each others' activities? Why is there such an absence of landmark discoveries which should give shape and coherence to any study of the development of knowledge? The sheer versatility of the leading scholars – typically mastering mathematics, astronomy and medicine, and still finding time to translate from several languages, and contribute to philosophy and poetry – suggests an imperfect commitment to any one pathway of knowledge. The dominance of the Greek authorities – or Indian in the case of mathematics – was so pervasive that the entire intellectual tradition sometimes appears as an extended commentary on Greek models. Yet if brilliant discoveries are missing, one is left with an overwhelming sense of the depth and vitality of scientific practice in Islam, of a culture deeply engaged with scientific thought in a way that was far in advance of the Christian west. Christendom's intellectual energy was poured above all into theological argument, into biblical exposition or ecclesiastical law. Of course Islam had its theologians and jurists, but the striking thing is that a strong tradition of rational and secular thought was able to flourish in this culture, whose very identity was supposedly founded on an imperious religion. This openness to the wisdom of other cultures was present in the earliest generation of Islamic thinkers, in Abu'Mashar, al-Kindi, and al-Razi, and reached its culmination in the work of the historian Ibn Khaldun (AD 1332–1406) who attempted to create a science of history. Ibn Khaldun saw that the history of human society is a process of organic growth and decay beyond conscious control, but what survives from one age to the next are the seeds of civilization, and in this view Islamic culture was but one phase in a wider pattern.

For many thinkers in the Islamic world, religious truth alone was clearly not enough. That the universe was under divine rule did not close off the search to understand the mechanisms of that rule. The central question which Islamic science provokes is why this intellectual activity should flourish in that particular culture: did the Islamic religion itself somehow release or inspire this scientific quest? There is obviously no final answer to a question like that, but the clue may lie in the character of Islam as an aggressive religion, a movement of conquest and domination: is it possible that knowledge itself was also seen as a realm to be conquered? Religions like Christianity and Buddhism centred on personal salvation, with this world seen as a temporary resting-place for mankind. But Islam was from the first a state and a law as well as a religion; it was all-embracing, and just as the Arab armies of the eighth century swept over much of the known world, so Arab philosophers of the ninth century seized the wisdom of the known world and made it their own. The Koran strongly emphasised design, order and purpose in nature, with man as the central and noblest part of creation. This was surely a sympathetic context in which science might be pursued; but the material of science, the data, was lacking, so having created an empire of the sword, Moslems set out to forge an empire of the mind, drawn from the intellectual riches of all their subject peoples. This they achieved, and Islamic science flourished, and fulfilled a vital historical role as a bridge between the classical world and early western science.

PART 2. CHRISTIAN PUPILS

&❧

'Medicine makes people ill, mathematics makes them sad, and theology makes them sinful'.

Martin Luther

THE DARK AGES

The years between 500 and 1000 AD have long been known as the Dark Ages of European history, with some justification, for during this long period of migration, war and the forging of small kingdoms, only the faintest traces of classical learning were preserved in the Christian monasteries. The one clear legacy of the Greeks – logic and metaphysics – was put to use in the field of theology, in articulating theories of the nature of God and man, employing such philosophical categories as substance, nature, form and quality. In such a culture, natural philosophy was in eclipse, and the history of scientific thought in this period is a fragmentary story of a few exceptional individuals and a few significant texts.

The outstanding intellectual event was the renaissance of learning associated with the court of Charlemagne in the eighth and ninth centuries, but even before that the figure of Bede in England (672–735) commands our attention. Spending his whole life in study and teaching at the monastery of Jarrow in Northumbria, Bede wrote extensively on scriptural and historical subjects, but also on the important problem of the measurement of time. The migration and intermingling of peoples, with different units of time, coupled with the importance of time in the religious life, elevated time-keeping into a serious art, known as *computus*. In his work *De temporum ratione*, Bede set out the astronomical basis of calendar calculation, especially the all-important methods for determining the date of Easter, a date which was not fixed because it depended on that of the Jewish feast of Passover, and the Jewish religious calendar was a lunar one. The question of the date of Easter had long been a vexatious one, and it was Bede's influence which fixed the formula of the full moon following the vernal equinox. He wrote a manual of basic astronomy to underpin his work on time, and described the classical constellations. Bede was the first to use the era of Christ's birth as a fundamental anchor date of history, a practice which spread swiftly throughout Christendom. Echoing Lucretius's title *De natura rerum,* and copying from Pliny and others, Bede composed a general descriptive text on the natural world linked to the hexameron, the six days of creation.

Bede was an isolated genius, but the court of Charlemagne fostered a host of scholars and artists. The concept of a Carolingian Renaissance is far from being a modern invention. 'Our times are transformed into the civilization of antiquity: golden Rome is reborn and restored anew to the world' boasted a court poet of the ninth century. Charlemagne inaugurated a deliberate process of educational reform, not only at his court but throughout his realms, importing scholars to staff cathedral and monastery schools. Alcuin of York, the foremost scholar in England in the generation after Bede, was invited to supervise the whole enterprise.

Easter-dating wheel. The problem of dating Easter greatly exercised the Church, and this twelfth-century manuscript shows an ingenious calculating wheel of 19 spokes representing the 19-year cycle within which the lunar and solar years coincide.
St John's College, Oxford

Among the classical works which were copied and used as textbooks were a number with a bearing on science, most notably the *Phaenomena* of Aratus, which served as an elementary survey of astronomy, with pictures of the constellations and celestial spheres. Charlemagne is known to have had a special interest in this subject, for his biographer Einhard writes that 'he investigated with great diligence and curiosity the course of the stars'. The astronomy of Aratus's text is descriptive and superficial, but it served to transmit a degree of astronomical knowledge in a non-scientific age. Alcuin was consciously employing the legacy of Greece to enrich Christian culture, to create 'an Athens more beautiful than the the ancient one, for ennobled by the teaching of Christ, ours will surpass the wisdom of the Academy.' Alcuin's ambitions were unfulfilled as far as science was concerned, for the Carolingian era was devoid of scientific thought, yet its educational reforms provided a vital environment in which scholarship began to flourish again.

One outstanding beneficiary was Gerbert of Aurillac (*c.*945–1003), who became Pope Sylvester II, an intriguing transitional figure who was one of the first Europeans to glimpse the richness of Islamic learning. Having studied at the monastic school at Aurillac, he crossed the Pyrenees to Vich in Catalonia where, influenced by the proximity to Islamic sources, he achieved a mastery of mathematics and astronomy that was in advance of any other scholar in western Europe. Gerbert was appointed head of the Cathedral school of Rheims, which he transformed into an intellectual centre that drew future kings and bishops to study there. Gerbert devised and wrote about a novel form of abacus which combined elements of Roman numeration with the Hindu-Arabic number-symbols, and which permitted multiplication and long division. He was a skilled musician and may have designed organs. He built sundials, and described the making of a spherical model with the celestial circles and constellations marked on it – such models were used by Islamic scientists but were unknown in Europe at this time – and Gerbert was probably the author of the earliest Latin treatise on the astrolabe, a work clearly copied from Arabic models. Under the patronage of Otto III, Gerbert was created Pope in the year 999, and dreamed of reviving the Carolingian Renaissance, but the young emperor died in 1002 and Gerbert himself the following year. Such was the impression he made on his contemporaries that legends almost immediately began to circulate attributing his extraordinary intellect to the devil's coaching, or to a magical head which answered questions for him. His career, with its renewed interest in the exact sciences and its exploratory contacts with the Islamic world, foreshadowed the future course of European thought.

Whether or not as a direct result of foundations laid in the Carolingian era, the twelfth century saw a quickening of intellectual life in the schools of France, Germany, England and Italy, schools which would ultimately grow into the universities of the thirteenth century. This was not an isolated phenomenon, but was undoubtedly related to the enrichment of urban culture which western Europe now experienced. There was greater stability of political boundaries, and some significant technological changes such as the use of the water-wheel and the invention of the horse-collar. Food supply, population levels and prosperity increased, and medieval Europe began to acquire its own distinct structures and identity. It would be wrong to pretend that science as we understand it now came into focus, but the medieval world-view became both broader and sharper. Theology still remained central to all intellectual endeavour, but Christian thinkers were increasingly driven by two concerns: the first was to demonstrate rationally the foundations of human knowledge, and especially knowledge of the divine; and second to demonstrate that the universe in which man found himself was a rational, ordered creation, which would in turn give proof of the nature of its creator. For these reasons it became essential to study and attempt to understand the natural world and the structure of the cosmos. The biblical account of creation

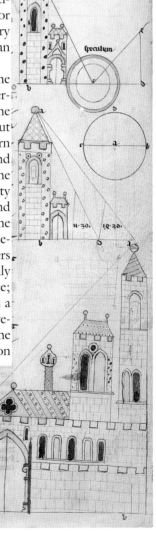

Measuring the height of towers in medieval geometry: three methods are given, all using the properties of triangles to find the unknown quantity. These were standard mathematical exercises, although a further method using the tower's shadow is not shown.

Bodleian Library, Oxford

presents us with a *fait accompli*, but what structures or mechanisms did the creator actually employ, and were they capable of rational analysis? The only available

guides in this quest were the classical philosophers such as Plato and Aristotle, who had devised certain explanatory concepts such as the celestial spheres and the four elements, and a number of more abstract formulae such as substance, form and purpose – an intellectual framework which was not to be found in the Bible itself of course. Thus from the twelfth century onwards there began a process by which western thinkers sought to reconcile the bare Genesis narrative with classical science to create a coherent philosophy of nature which could exist in harmony with Christian revelation.

The creation of the heavens, a fourteenth-century illustration to the Book of Genesis. Medieval science saw its central task as rationalizing God's creative activity.
The British Library
Egerton Ms. 1895 f.5v

This was a profound enterprise and one which would lead to the creation of an ingenious and lucid medieval world-view; but it should be emphasized that medieval intellectual discourse was overwhelmingly book-centred, and dominated by the idea of authority: ideas were derived from texts, and texts were weighed against other texts. Science, like all other study, meant learning rather than discovery. The twin sources of knowledge were revelation and reason, there was no other, and the authorities of the past owed their pre-eminence to the power of their reasoning. There was, in a word, no experimental or empirical approach to nature. Astronomy was what Ptolemy had reasoned it to be; natural history was what Pliny had observed; medicine was what Galen had described. Medieval thinkers sought, and achieved, a coherent natural philosophy, but it would be one that would eventually and inevitably crumble under the impact of an empirical revolution.

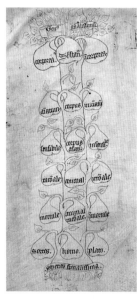

The Tree of Porphyry typifies the medieval practice of organizing knowledge into hierarchical schemes and structures, often visualized as picturesque trees or wheels. In this tree, the species man at the base is progressively defined as animal, mortal, rational and sensible, and these qualities constitute his unique substance in a world of substances.
The British Library
Royal Ms. 8.A.XVIII.f.3v.

THE RECOVERY OF GREEK AND ISLAMIC SCIENCE

In this enterprise, the recovery of Greek science was vital, and the principal pathway through which that science reached Europe was Islamic Spain. Gerbert had pioneered the scholarly discovery of the riches of Islam, and there were other tenuous contacts too, such as an exchange of ambassadors between the courts of Otto the Great and the Emirate of Cordoba around the year 950. The most sustained occasion of contact between Christian and Muslim was of course the crusades, but these had little intellectual impact, presumably because those engaged in them were warriors and not scholars. The vital event was the Christian reconquest of Spain during the eleventh century, when centres of Arab culture came under Christian control, together with educated Mozarabs – Christians who had been allowed to practise their religion under Moorish rule – able to translate Arabic texts into Latin. A single event, the fall of Toledo in 1085, placed the most extensive philosophical and scientific library in Europe at the disposal of western scholars, and a movement to translate its rich possessions soon began. Some of the translators were Spaniards, others came from Italy, Germany and Britain, perhaps the earliest and among the most important being Adelard of Bath (*fl.*1116–1142), who translated some of the astrological work of Abu'-

Mashar, and the astronomical tables of al-Khwarizmi, thus giving western scholars a taste of both the excitement of Islamic speculation and the precision of its science. He also produced a seminal Latin version of Euclid, translated from Arabic, which replaced the faulty version then current in the west. If Adelard was the earliest intermediary between Islamic learning and the west, undoubtedly the most prolific was Gerard of Cremona (*c*.1114–1187) whose versions of almost a hundred scientific works made him the most important scientific writer of his era. Among the Greek works which Gerard translated from Arabic were those of Aristotle, Euclid, Archimedes, Ptolemy and Galen, while the Islamic scientists whose thought he made available to the west included al-Khwarizmi, al-Kindi, al-Razi, Ibn Sina, al-Farghani and Jabir Ibn Hayyan. Gerard is known to have employed assistants and collaborators, and the whole Toledo enterprise is reminiscent of that in Baghdad three centuries earlier – the deliberate assimilation of an older and stronger intellectual tradition.

Although Spain was the setting for the most intense phase of this translating activity, and Arabic the vital intermediary, the twelfth century also saw some important translations direct from Greek into Latin by James of Venice

(*fl*.1136–1148), including Ptolemy's *Almagest* and Euclid's *Elements*. Italy, with its contacts with the Byzantine world, was the natural location for translation from the Greek, yet it was also, through the figure of Leonardo Fibonacci, the setting for the introduction of a revolution in mathematical methods from Islamic North Africa. Fibonacci (*c*.1170–1240, also known as Leonardo of Pisa) travelled in Sicily, Algeria, Egypt and Syria, where he was deeply impressed by calculation methods used in commece. Returning to Pisa, he taught and wrote on these methods, and his work *Liber Abbaci* was the first western mathematical treatise to make systematic use of the Arabic, i.e. Indian, numerals, explaining the concept of place-value to determine whether a given number represents units,

Medieval scientists observing the heavens, using an armillary, a square and dividers. The men are said to be Egyptians and the text repeats the classical and medieval belief that mathematics and geometry were invented by the Egyptians.
The British Library
Royal Ms. 20.B.XX. f.3r

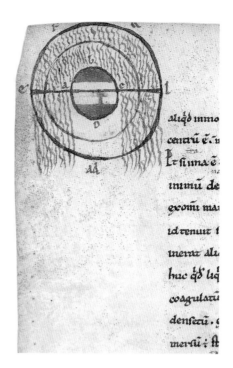

The sphericity of the earth. The first rather crude fourteenth-century picture illustrates a solar eclipse, but is more interesting for its clear demonstration of the sphericity of the earth; the 'flat earth doctrine' of the middle ages is a persistent modern myth. The other picture shows what would happen if the spherical earth were not the centre of the universe to which all falling bodies descend, as it is in Aristotle's physics: rainfall, instead of falling evenly over the whole surface, would miss part of the earth and continue out into the celestial sphere, an absurd supposition.
The British Library
Royal Ms. 19.C.1.f.40v (left) and Harley Ms. 4794 f.46v (right).

tens, hundreds or thousands. The word *abacus* in the title does not refer to the traditional instrument, but means computation in general. The Arabic numeral system was known only to a few European specialists who had studied Gerbert or who had access to Gerard's more recent translation of al-Khwarizmi; by demonstrating its use in practical problem-solving, Leonardo ensured its widespread adoption. He applied the new notation to questions of profit, interest, proportions and unequal shares, and the conversion of money and of weights and measures. The flexible and powerful properties of the new notation system inaugurated a new era in European mathematics, and had an immense impact on commerce, especially facilitating the growth of credit and banking. On the more speculative side of mathematics, Fibonacci's name is preserved in the Fibonacci sequence 1, 2, 3, 5, 8, 13, 21, 34 and so on, where each number is the sum of the two preceding numbers, a sequence which he was the first to describe.

Much of this activity of assimilation from Arabic or Greek sources was exclusively of a secondary, literary nature; it was not original scientific thought or research as we would understand it, or even as the Islamic masters had practised it. Yet the striking thing was that the scholars of Christian Europe were able to recognize immediately the superiority of Greek and Islamic science, and that they felt it imperative to revitalize western culture by opening it to these new influences; to renew its mastery of the exact sciences of mathematics and astronomy, to strengthen its grasp of the philosophical bases of science, and to introduce new approaches to the natural sciences of medicine and astrology.

Although the origins of this new learning lay in pagan and Islamic science, its overall impact was to foster a new naturalism in medieval thought that was welcomed as theologically satisfying by the Christian west. Plato's creative principle, the craftsman, was readily identified with God, and the complexities of the natural world were seen as self-sustaining mechanisms built into its structure – although these were always capable of being overruled by the creator, as happened in the miracles performed by Christ or by the saints. Thus the idea of the universe as the creation of a personal and beneficent God was potentially favourable to the natural sciences, in contrast to the animism of pagan religion, in which the features of the earth and the heavens – rivers, seas, storms, wild animals, the Sun and Moon – were under the capricious control of a host of petty and perhaps demonic spirits. The Christian, and indeed the Islamic, God was conceived to have shaped the world through the laws of natural causation, as the elements and forces of the universe interact upon each other. This principle was well expressed by the great teacher Thierry of the cathedral school of Chartres in the twelfth century, renowned in his own day as an expert on mathematics and classical philosophy. Thierry embodied his natural philosophy in an exposition on the first chapters of Genesis, in which he taught that God's initial act of creation was to devise the four elements in a single instant, and that the form which the world subsequently assumed was the result of natural processes. Fire, once created and unable to remain still, immediately began to rotate as a 'fiery heaven' or Sun; the heat evaporated the water, causing it to ascend as vapour to form the 'waters above the sky'; this removal of water left dry land on the earth. The heating of the waters above the sky formed the stars, planets and Moon, which are composed of water, while the heating of the land engendered the living forms of plant and animal life, among them man and woman. This was Thierry's rational understanding of the six days of creation – events unfolding through laws inherent within the physics of the four elements. Crude and imprecise as a sequence of events, it nevertheless treats nature as an autonomous power, to which Plato and Aristotle are clearer guides than the Bible. This approach became for a time an accepted doctrine among the schoolmen, but such an invasion of religion by the forces of secular philosophy would inevitably provoke a reaction, just as it did in the Islamic world, and that reaction came when Aristotelianism seemed to the orthodox to be threatening to overwhelm biblical truth.

THE ARISTOTELIAN REVIVAL

Plato's science, through Latin translations of *Timaeus,* had always remained accessible to the west, and its central doctrine of design in the universe was easily re-interpreted in a religious light. Aristotle's works however were, with a few exceptions, not part of the west's inheritance, and their rediscovery came via translations from the Arabic in the twelfth century, and equally from the commentaries on Aristotle by Islamic scholars, especially Averroes. In the thirteenth century a fresh wave of translations direct from the Greek flowed from the pen of William of Moerbeke (*fl.*1260–1286), among them several of Aristotle's works never before available in the west. Aristotle's thought offered a persuasive conceptual model for understanding the natural world, one that was more precise and analytical than Plato's, and which was eagerly accepted by Christian philosophers as an essential framework within which to place religious truth. This intellectual synthesis was developed in the twelfth and thirteenth centuries by the foremost thinkers in Christendom: Robert Grosseteste, Roger Bacon, Albertus Magnus and, most famously, Thomas Aquinas. Grosseteste (*c.*1168–1253) was Chancellor of Oxford University and subsequently Bishop of Lincoln, a leading figure in English intellectual and ecclesiastical life. A convinced Aristotelian, he advocated a systematic approach to problem-solving, proceeding from 'the fact' (*quia*) to the 'the reason for the fact' (*propter quid).* Mathematics he considered to be central to the analysis of reality, arguing that 'all causes of natural effects have to be expressed by means of lines, angles and figures, for otherwise it would be impossible to have knowledge *propter quid* concerning them'. Yet combined with this analytical approach was a strong current of Neoplatonic mysticism, seen in his belief that light was a fundamental form of nature, that it was created first of all forms and propagated itself from a single point into a sphere, thus defining all spatial dimensions. This diffusion of light he believed must occur in strict mathematical proportions. Light was more than the familiar visible light, but was considered to be a creative medium varying in its qualities and effects, sunlight for example being responsible for the transformation of one element into another, while it was moonlight rather than the moon's mass which caused the tides. Grosseteste even proposed that light was the key to astrological influences, for the light of the heavenly bodies reached the earth in complex geometric patterns reflecting the aspect of the heavens, and it was this infinitely varying pattern

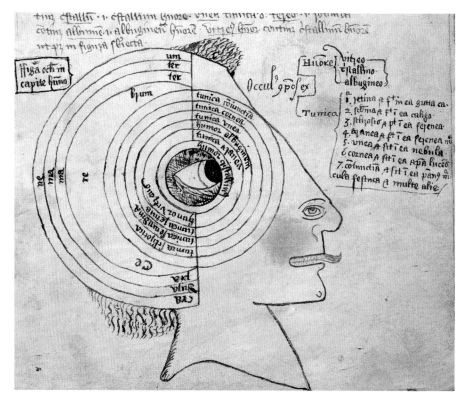

The parts of the eye, from an anonymous fourteenth-century manuscript. The concentric circles are not intended to represent the eye's true structure, but are labelled with the names of the fluids and membranes within the eye – the cornea, the lens, the retina and so on. These parts had been identified with surprising precision.
The British Library
Sloane Ms. 981 f.68r.

which affected human character and events. In the investigation of light he made many experiments with optics, and plainly understood the laws of perspective and their effects, that 'the size, position and arrangement according to which a thing is seen depends on the angle through which it is seen, and the position and arrangement of the rays, and that a thing is made invisible not by great distance, except by accident, but by the smallness of the angle of vision.' He was an expert in *computus,* and drew attention to the imperfections of the Julian calendar – three centuries before it was reformed. Like other medieval scientists, Grosseteste's experiments were overshadowed by theory and metaphysics, but his faith in nature's mathematical unity was of lasting interest to philosophers, and his works were much re-printed in Venice and Nuremberg in the sixteenth century.

The successor to Grosseteste in England was the celebrated Roger Bacon (*c.*1219–*c.*1292), perhaps the only medieval scientist to enjoy a popular reputation

– partly as a magician and partly as a prophet of new technology. His career began as a conventional philosopher, but around 1247 he conceived a highly unusual interest in science, acquiring, in his own words, 'secret books', instruments and experimental equipment, which he used in Oxford and Paris, where he taught. His actual achievements through experiment were meagre, but his works are outstanding in their time for their explicit advocacy of experience and reason and for their criticism of the authorities of the past. He expresses complete confidence that a knowledge of nature's richness will lead to a deeper knowledge of God. He echoes Grosseteste's conviction that light, as well as sound and heat, are propagated according to mathematical laws – he uses the phrase *lex nature universalis* – and optics retained its central place as a subject of study. He studied the behaviour of lenses and of magnets, and proposes dietary experiments with animals to discover which foods will prolong human life. He describes ingenious machines for flying, lifting weights, submarine travel, which he believed had been built in antiquity and which could be built again. It is on these works of imagination, and on his directions for making gunpowder, that his reputation as a proto-Leonardo rests, but we have no evidence that he attempted to build any of these devices. It is certain however that he experimented with lenses, and it is just possible that he made some early form of telescope, for he wrote of 'reading the smallest letters from an incredible distance' and of 'causing the Sun, Moon and stars to descend in appearance here below'. His sense of scientific geography was in advance of his contemporaries, for he advocated the use of latitude and longitude at a time when it was unknown in European mapmaking. He discussed the sphericity of the earth and the possibility of sailing west from Spain to Asia (this part of his work was known to Pierre d'Ailly, a source of inspiration to Columbus) and is known to have compiled world maps, now lost. He investigated the concept of the mechanical clock, although without apparently building one, suggesting that if the face of the astrolabe might be turned at a constant rate, it would function as an analogue of celestial motion.

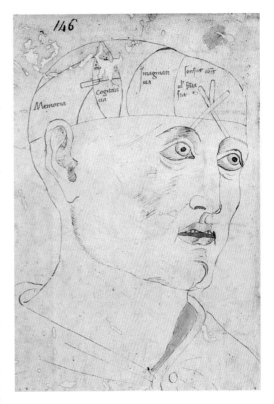

Bacon's was a restless intellect, born out of its time; inevitably his achievement failed to match his ideals, but his approach was radical and visionary. Conscious of his own originality, Bacon was undisguisedly jealous of the most renowned philosopher-scientist in Christendom, Albertus Magnus (*c.*1200–1280), who played a large part in introducing Greek and Arabic science into European universities. The newly-translated works of Aristotle required elucidation, and this is what Albertus provided, issuing commentaries on every known Aristotelian text, and in the process providing a summary of all contemporary knowledge. Throughout this vast enterprise he advocated the testing of received teaching by experiment, a term which meant for Albertus observing and analysing phenomena for oneself. He was not afraid to disagree with Aristotle on matters of detail, or on fundamental principles, such as the eternity of the universe, which he regarded as philosophically absurd. He studied the medieval core-sciences such as astronomy and optics, but his interest in the earth- and life-sciences was outstanding in its time, and he described animals, plants and minerals, classifying them according to their forms. Alert to the possibility of conflict between faith and

A medieval map of the brain. Aristotle had claimed the heart to be the seat of the soul, whereas the medical tradition of Galen placed the soul in the brain. Here the brain is divided into four cells controlling, from front to rear, common sense, imagination, reason and memory.
The British Library
Add.Ms.22553 f.1v.

reason, he argued that all that happens does so by God's will; it is the role of science to investigate the mechanisms employed by the creator, the classic example used in this context being that of the flood, that when God wished to punish the wicked, he did so by invoking the destructive forces of nature. This rational philosophy of nature was absorbed by Albertus's most famous pupil, Thomas

Table of the hours of moonlight, shown inside the outer circle, directly related to the age of the Moon in days, seen inside the central trefoil.

St John's College, Oxford

Aquinas, who was not a scientist, but who was anxious, for religious reasons, to demonstrate that the universe was an ordered system whose laws might be studied, and at least partly understood. His theological works touch on subjects such as motion, gravity, the celestial bodies and the four elements, and on what he termed the 'occult workings of nature', such as magnetism and tides, where his concern was to argue that such things do indeed have natural causes which will one day be better understood. Likewise he regards the Ptolemaic system as an interim report and an excessively complex one, which will be replaced by a simpler and more accurate account.

The great lesson which all these Christian philosophers had drawn from Aristotle was the centrality of reason: reason alone had led Aristotle – a pre-Christian, pagan thinker – to a true knowledge of the soul, of the structure of the universe and of the divine power who moves it. Therefore there *is* a natural knowledge of God, a rational faculty in the human mind capable of analysing the natural world and deducing the character of its creator. There was of course another source of knowledge, namely divine revelation, recorded in the Bible; but it was impossible for revelation and reason to be in conflict, for as Aquinas argued 'What is divinely taught cannot be contrary to what we are endowed with by nature: one or the other would have to be false, and since we have both from God, he would be the cause of our error, which is impossible.' This principle functioned as a charter for natural philosophy, in which reason was free to seek out evidence of design in the universe as a key to the mind of its creator.

Nevertheless many in the Christian church were still deeply uneasy at this philosophical invasion, and they pointed out vigorously that Aristotle's thought contained a number of central ideas which were irreconcilable with the teachings of the Bible. One such obvious conflict was Aristotle's conviction that the universe was eternal, without beginning or end. Another was the doctrine that the soul was the form of the body, inseparable from it, flatly contradicting the belief in the soul's immortality after bodily death. And perhaps the crucial ground for hostility to Aristotelianism – and indeed to the whole tendency towards naturalism in medieval thought – was the problem of determinism: if the universe operates through natural cause and effect, what role is left for God and his con-

trolling power? To assert that natural causation accounted for the forms and forces of nature, carried with it an alarming implication of determinism, which could not be reconciled with the omnipotence of God, or with human free will.

Angels moving the heavenly spheres. Aristotle had attributed the movement of each heavenly sphere to a *primum mobile,* a form of divine entity. This was unacceptable in Cristian thought, and the task was transferred to angels, seen here turning crank-handles.
The British Library
Royal Ms. 19c.I,f.34v.

These were serious problems which illustrate the barrier to the pursuit of natural philosophy within the context of this deeply religious culture: if nature is an autonomous system, then the powers of the deity were felt to be compromised; yet if nature is under the arbitrary control of a deity, can the laws of nature be said to exist? Is nature sometimes a consistent, autonomous system, and sometimes not? Suspicion of the rational, scientific enterprise resulted in the issuing in 1277, three years after Aquinas's death, of an edict by the Bishop of Paris condemning as heretical several hundred philosophical propositions, many drawn directly from Aquinas's work. This was not a decree binding on the church at large, for it related only to the content of teaching at the university of Paris, but its importance lay in the fact that Paris was at this time the undisputed centre of Christian theological activity. Central among the condemned teachings was the doctrine that the processes of nature, its 'secondary causes', were autonomous, and would continue to function in the absence of the 'primary cause', namely God. The great concern was that Aristotelianism, rationalism, and philosophy generally, was seeking to restrict God's freedom through the development of a secular natural philosophy, a position which would be intolerable to the church if stated in those terms. Are this world and its laws dependent on the arbitrary will of God, and if so, might God have created a different kind of world? Or is there some inner necessity for the world to function as it does, again implying a restraint on God's freedom? It was on this, the metaphysical context of science, rather than on science itself, that so much intellectual energy was expended.

What was the long-term result of the condemnation of 1277, and what specific developments were there in western scientific thought in the thirteenth and fourteenth centuries? It seems that, despite the efforts of the Parisian authorities, Aristotelianism was was proving irresistible to Christian philosophers, for by 1323 Aquinas was canonised, with the additional title 'Doctor of the Church', and considerable energy was being devoted, mainly in the universities of France, England, Italy and Germany, to building on the classical foundations – as mediated by Arabic scholarship – in astronomy, physics, optics, mathematics and medicine.

THE PHYSICAL SCIENCES

At the heart of medieval science lay the study of astronomy. It addressed the profoundest problem of all, namely the structure of the universe, and its field of study -heaven – was not only a physical place but a spiritual dimension, inhabited by God and his angels. Astronomy therefore interacted with philosophy and with religious belief, and its intellectual implications were greater than those of any other discipline. Western astronomy attained some degree of maturity during and after the twelfth century when, under the tutelage of Islamic science, the legacy of Greece was recovered. There was fundamental agreement that the heavens must be understood in terms of spheres and circular motion with the earth at the centre, but beyond that, medieval astronomy, in Christian and Islamic thought, was very far from being monolithic. In the first place, there is a useful distinction to be made between mathematical astronomy, the precise plotting of celestial movements and positions, and cosmology, which attempted to get behind the observational data and theorize about the structure and functioning of the universe. The great master of mathematical astronomy was Ptolemy, whose complex geometry of cycles, epicycles and eccentrics aimed at providing a systematic description of the movements seen in the heavens, but which was silent on the physical forces which could underlie those movements. Moreover the dominant, orthodox physics of Aristotle envisaged the cosmos in terms of uniform circular motion, as a cluster of concentric spheres, which was impossible to reconcile with the complexities of Ptolemaic geometry. This fundamental conflict generated innumerable theories and systems, nor was there ever complete agreement even about the number of the heavenly spheres, or the order in which they encircled the earth. Some astronomers argued that, while the Sun, Moon and outer planets orbited the earth, Mars and Venus actually orbited the Sun, thus accounting for their constant proximity to it. It was doubted that the astral sphere could be the outermost sphere, since the Bible referred to the 'waters above the heavens', so a further sphere of crystalline, frozen water was added; but then the doctrine of the four elements gave priority to fire as the lightest element, and it seemed inevitable that a sphere of fire, the empyrean, must be the final sphere of all – although some authorities dismissed them both as fables.

One notable attempt to anchor the Ptolemaic system in physical reality was made by Campanus of Novara around the year 1260. Beginning with the earth-moon distance given in the *Almagest,* Campanus set out to calculate the dimensions of the cosmos, of each cycle and epicycle, as far as the sphere of the fixed stars. He did this by adhering to the Aristotelian doctrine that there is no empty space in the universe, and that the farthest distance from earth reached by one planet is the nearest reached by the next. The general order of Campanus's results are similar to Ptolemy's, worked out with scrupulous, indeed impossible accuracy. Campanus's scheme fixed the classical dimensions of the universe for the next three centuries, until the re-evaluation forced by Copernicus. Campanus was also the first western scholar to describe the instrument known as the equatorium, a plane model of the planetary cycle and epicycle, set within the ecliptic, which permits the finding of planetary positions just as the astrolabe does stellar positions.

The nature of the spheres themselves, and the force which moved them, was among the most mysterious of problems, for in Aristotelian physics, movement

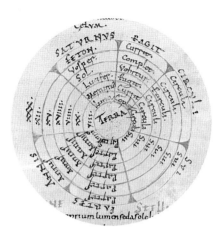
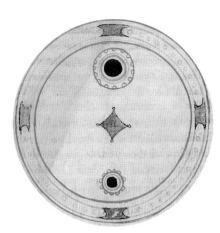

Left: Wheel-diagram showing the period of rotation of the planets, working inwards from Saturn, with its period of 30 years. From a tenth-century manuscript of Isidore of Seville.
Bibliothèque Nationale, Paris

Right: Sunspots in a twelfth-century manuscript. In a monastic chronicle, John of Worcester records that on 8 December 1128, two black shapes were to be seen on the Sun's disk, one in the upper and the other on the lower part, as shown in this remarkable drawing.
Corpus Christi College, Oxford

occurs only when it is caused, continuously caused by a continuous force. Aristotle had assigned a prime mover, an intelligence of some sort, to each celestial sphere, but this was unacceptable in a monotheistic religion, unless it was the angels who turned the spheres – and in a number of medieval illustrations, angels are indeed shown turning the spheres with cranks. Far more sophisticated was the concept of the Parisian scholar Jean Buridan, who suggested that the heavens had been set in motion by God at the creation, and that the impetus of this act was still effective, thus sweeping away the troubling crowd of celestial intelligences, demigods or angels. The notion that the sphere was nature's ideal form was extended also to the earth, and it is beyond doubt that medieval thinkers were aware of the sphericity of the earth; the flat-earth legend has grown up through a misinterpretation of medieval world maps – that their circular, disc-like shape signified a belief that the earth was indeed a flat disc, whereas this was merely a graphic convention. By the fourteenth century, medieval cosmology had moved far from its rigid, stereotyped image, and could entertain some quite radical theories. For example Buridan and his pupil Nicole Oresme noticed how neatly the diurnal movements of all the heavenly bodies might be explained by the rotation of the earth itself, rather than by the system of multiple spheres. But neither the physics nor the common-sense of the day could assent to a moving earth, and this remained only a speculation.

When western scholars awoke to the possibilities and the demands of mathematical astronomy, they had of course no body of observational data from which to build models of celestial movements or predict future positions. These data they had to borrow from Arab sources, and the translation of astronomical tables, the *zijes*, like that of al-Khwarizmi, placed an essential tool in their hands. The first new tables compiled in the west were published under the patronage of Alfonso X of Castile, for which new instruments and new observations were used to produce a source-book for astronomy dated to the epoch 1252. The astrolabe and Latin treatises describing its use were also modelled directly on Arabic sources. Alongside these tables and treatises appeared the first western textbooks of

Portrait of Nicole
Oresme, the fourteenth-
century mathematician
who speculated that the
diurnal movement of the
heavens could be elegant-
ly explained by the rota-
tion of the earth itself.
Here Oresme presents his
Livre du Ciel to King
Charles V of France.
Bibliothèque Nationale, Paris

astronomy, which conveniently summarize the level of astronomical knowledge
in the universities. The most popular were two works both entitled *De Sphaera*,
one by John of Hollywood, or Sacrobosco, and other by Campanus of Novara.
They circulated in hundreds of manuscript copies, and instructed the reader in the
elements of Ptolemaic, spherical astronomy, in the necessary calculations, and in
the art of *computus* – time-reckoning.

The most powerful motive behind the western awakening to astronomy was
without doubt its astrological dimension. Condemned by the church fathers in
the post-classical era, astrology had
formed no part of the western
intellectual tradition until important source-books like Ptolemy's
Tetrabiblos and Abu'Mashar's *Introduction* were translated into Latin
in the twelfth century, and the
astrological revival began. It found
fertile soil in the Christian world as
it had in the Islamic, because it
reinforced the mystical tendency of
these religions, the desire to make
comprehensible the soul's sense of
being part of the cosmos, and
standing in a numinous relationship to its creator. As a description
of subtle but still concrete *physical*
forces believed to operate in the
cosmos, astrology became an
accepted part of natural philosophy. Ptolemy, Abu'Mashar and all
astrological authorities were convinced that some physical power

The personified figure
of Saturn in his chariot,
from an early printed
Latin edition of
Albumasar's
(Abu'Mashar) *Book of
Conjunctions,* one of the
fundamental source-
books of medieval
astrology.
The British Library IA 6690

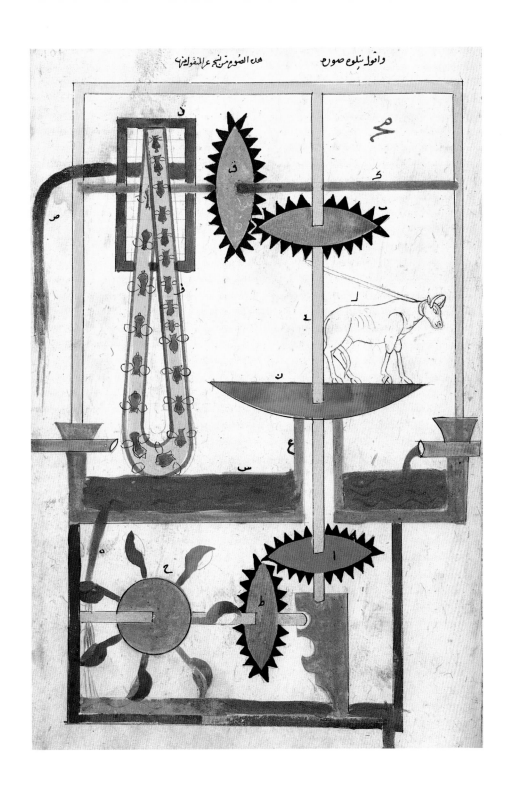

وقالوا بلوغ صوره هذه الصوره منسوخة عن المغولينه

In thirteenth-century Persia, al-Jazari emulated the mechanical inventiveness
of Hero of Alexandria, designing automata driven by water, heat or falling weights.
Bodleian Library, Oxford

God as geometer of the universe. The rational theology of the middle ages held the central principle that there was no conflict between revelation and reason, for both were the work of God. This finds its symbolic expression in the image of God creating the cosmos in accordance with mathematical laws, measuring it with a pair of compasses.
Österreichisches National Bibliothek, Vienna

The constellation Cepheus, from a magnificent ninth-century Carolingian manuscript of Aratus's *Phaenomena*. Aratus's description of the constellations served as a source-book of basic descriptive astronomy throughout the middle ages.
Leiden University Library

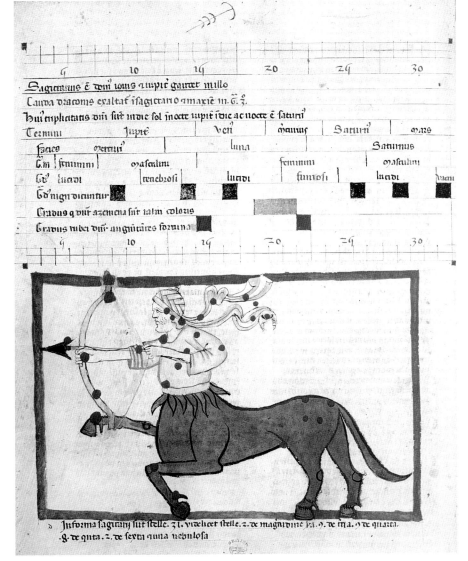

Precision in astrology: this fourteenth-century picture of the constellation Sagittarius is divided into thirty degrees, within which all the astrological information is tabulated: which planets rule in each degree, which are feminine or masculine, and so on. The constellation itself is rather precisely drawn, with 31 constituent stars graded for magnitude.
The British Library
Add.Ms.23770 f.18v.

emanated from the heavenly bodies, and produced effects on the elements which composed the earth and the humours of the human body – in medicine, astrology became an essential diagnostic aid. The art of casting individual horoscopes was more controversial, and the church opposed the belief that human life could be determined by these forces, for it compromised both human free will and divine omnipotence. Astrology was among the new rationalist teachings condemned by the Paris authorities in 1277, along with the new Aristotelianism, but to little effect, for its allure was irresistible. Mathematical astronomy was essential to astrology in plotting the all-important positions and conjunctions of the heavenly bodies. Astrology, with its principal service of casting horoscopes, made possible the pursuit of astronomy as a career, and drew many leading intellects to the discipline. The extent to which astrology permeated intellectual life by the fourteenth century may be seen in the works of Chaucer, for it has recently been demonstrated that many of his narrative poems have concealed astrological meanings: the characters and incidents in *The Canterbury Tales* and *Troilus and Criseyde* have been shown to correspond to movements among the stars and planets, and it has been conjectured that Chaucer used an astrolabe (on which he wrote an expert treatise) to reveal this hidden dimension, after reading the poems aloud to his audience.

Medieval astronomy rested on a long tradition of observation and mathematical analysis, and it achieved rational, coherent results – although many of its tenets were of course inaccurate by modern standards. By contrast medieval physics was less securely based on observation and measurement, and it tended to be so dominated by metaphysical concepts that it has the appearance of an alien landscape where few features are recognizable to us. Derived from the Greek word *physis,* meaning nature, its purpose was to investigate the nature of things in the material world, and its scope consequently included many areas which in modern terms fall within chemistry, biology or geology. The central focus of attention were the processes of change, changes in quality, changes in position, growth and decay. Yet medieval scientists were fatally hampered by their lack of observational technique, of instruments and accumulated data, and instead the problems of diversity and change in nature were approached via an edifice of metaphysical ideas derived largely from Greek authorities. The fundamental framework was made up from the Aristotelian notions of form and matter. Form is the essential character of an object, the sum of qualities which go to make up a tree, a horse or a stone, while matter is merely a neutral shell. Form and matter are of course always combined, and although they may be analysed and discussed separately, they never occur separately in nature; thus the greenness of a green leaf, or the heaviness of a rock cannot be separated from those objects, yet they are real qualities. The primary qualities of the four elements were considered to be such forms: heat, dryness, cold, and wetness formed respectively fire, air, earth and water, of which all matter was composed. Since matter is capable of assuming a succession of forms, it follows that the elements may be changed one into another. How this process of change occured was perhaps the great crux of medieval physics, and the difficulty of explaining it led the great Islamic philosophers Avicenna and Averroës to introduce a third category, which they called corporeal form, which acted as an intermediary between form and matter. Corporeal form was the organizing agent which gave specific shape and character to matter, to produce each object. We might think of it as a genetic code which organizes life into different organisms or species, but medieval philosophers thought of it as existing on a different plane of being, perhaps like the Platonic forms which could be understood as existing in the mind of God.

Theoretically, the doctrine of the four elements required that the earth be formed of four concentric spheres, the innermost being the heaviest element, earth, surrounded in ascending order by water air and fire. In reality these four elements had become mixed, perhaps through the agency of the Sun's heat and light. Exactly how the elements combined and separated to form all the substances found on earth, was one of the great unknowns of science, but nature was clearly conceived to be in a state of eternal flux, and in this respect no special distinction was made between the organic and the inorganic realms. One classical concept which held no appeal either for Islam or for the Christian west was atomism, with its impersonal determinism, uninfluenced by divine or human will. In Aristotelian thought, all material substance was reducable to minute building-blocks, termed *minima naturalia*. But where atoms, in the thought of Lucretius for example, were all identical in character, *minima* were still composed of their parent substance: however small they were, *minima* of iron, honey or gold were still iron, honey or gold.

The discipline founded on the idea of nature's unity, and which attempted to work out the realities of change, was alchemy. It had arrived in the west through the twelfth-century translations of Geber, and was discussed by philosophers like Bacon and Albertus Magnus, who regarded it optimistically as the genuine science of matter. Alchemy acquired a language and an apparatus of its own and had the appearance of a systematic science. It did indeed result in genuine discoveries, such as the mineral acids and alcohol, and alchemical treatises poured from the

Facing page: Combinations of elements: one of the great problems with the classical doctrine of the four elements was to understand how they might combine to form the myriad substances in the world. This table does not answer that problem, but it shows the scores of possible combinations, with first one then another element dominating the others.
The British Library
Harl.Ms. 4348 f.26v.

Left: Students of alchemy in the foreground, receiving instructions from the great masters of the art, including Geber, al-Razi and the legendary Hermes.
The British Library
Add.Ms. 10302 f.32v

pens of the practitioners, although they were almost invariably ascribed to the great masters of the past, such as Geber, and given titles like *The Sum of Perfection*. The ideal of transformation of the elements became concentrated on the production of gold and on the search for the elixir of life, and alchemy became overlaid with occultism and charlatanism. Figures such as Nicolas Flamel of Paris, who, in the 1360s, claimed he had dreamed of an occult book and subsequently found it and with its guidance had made gold, have contributed to the glamour of alchemy, but have made its place in the history of science highly questionable. By 1400 the same formulae were being prescribed for the ennoblement of metals as for the preservation of life, and the elixirs became medicines. Despite its occult vocabulary, alchemy is of great interest in the medieval context as a practical, empirical pursuit, an attempt to grapple with the substances of the material world. Inevitably its practice was compromised by the confusions of its analytical framework, and of its metaphysical guiding principles.

That empirical research and results were not inherently beyond the reach of medieval scientists is demonstrated in the work of Pierre de Maricourt, (*fl.*1269, also known as Peter Peregrinus) author of the first treatise on magnetism. Based on extensive observation and experiments with lodestones, Maricourt introduced the concept of north and south poles, considering that the attractive power resided in the celestial poles above the earth. He went on to describe the magnetic compass, how it might be constructed and the benefits that would flow from its use by mariners. This work is an example of experimental research and deduction almost unique in its time, and one whose results remained influential in the scientific era of the seventeenth century.

In the fourteenth century, in the universities of Oxford and Paris, important steps began to be taken towards the mathematization of physics. The problem of motion had long been discussed in metaphysical terms; for example all natural motion was supposed to express the tendency of all objects to seek their natural place in the universe, while forced or impressed motion occurred only in response to a constant motive-power. This supposition led to great difficulties in accounting for, on the one hand, the movement of the heavenly bodies, and on the other the flight of a spear after it had left the thrower's hand. Great philosophical energy was expended also in determining whether motion actually exists, or is it merely a succession of positions of the moving object? This idea goes back to the Greek logician Zeno and his assertion that the runner Achilles can never catch the tortoise, because when Achilles reaches the point where the tortoise stood, the animal will already have moved on: motion or speed are not real properties possessed by Achilles. This level of debate flourished among the metaphysicians, and it must be acknowledged that the material culture and technology of the middle ages did not encourage concepts such as mass and acceleration, or even the accurate measurement of speed or weight, volume being the common gauge of materials. By around 1350 however, a group of scholars at Merton College, Oxford had begun to look at systematic ways to analyse motion mathematically. It was seen too that other natural qualities such as heat, weight, force or speed might be treated as if they possessed extension, and therefore could be represented by lines in a geometric figure. The Merton group included Thomas Bradwardine, later Archbishop of Canterbury, and Richard Swineshead, and their work resulted in a new vocabulary of velocity and acceleration, expressed in formulae. One example of their work has become known as the 'Merton rule', or mean-speed theorem, which stated that an accelerating body covers the same distance in a given time as it would moving for that same time at its mean speed. This theorem was soon given elegant geometric proof by the Parisian Nicole

Oresme, in what amounts to coordinate geometry. Elementary as it now appears, it represents an important breakthrough in the application of mathematics to physics.

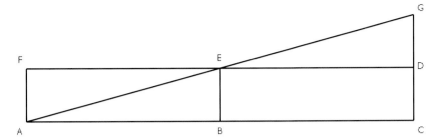

That mathematics was experiencing a revival generally is illustrated in the career of Richard of Wallingford (*c*.1292–1336), Abbot of St Albans. His work *Quadripartitum* was the first comprehensive western treatise on trigonometry, giving formulae for solving problems in spherical astronomy using sines and chords, and

The mean speed theorem, given geometric form by Nicole Oresme in an early application of mathematics to a problem in physics. ACG = uniform accelerated motion; BE = mean speed; ACDF = uniform motion; the area of ACG = the area of ACDF.

By courtesy of D.C. Lindberg

Richard of Wallingford, the fourteenth-century Abbot of St Albans, was one of the greatest mathematicians of his day, and was famous as the builder of one of the earliest mechanical clocks, driven by falling weights. Here he is seen measuring divisions on a circle. The leprosy which ended his life is plainly visible in this picture.

The British Library
Cotton Ms. Claudius
E.IV.f.201r.

was made possible entirely through the author's mastery of Islamic sources. Richard designed an improved equatorium for calculating planetary positions, called the albion – the 'all by one' – and achieved his greatest fame by building the first mechanical, weight-driven clock of which we know the details. As well as indicating the time, Richard's clock had additional trains for displaying the movements of the Sun, Moon and planets. The face was that of an astrolabe several metres across, and the movement of the pointers was conceived as an analogue of the movement of the heavens. The clock soon acquired great symbolic interest for scientists and theologians, as an image of the cosmos: complex, purposeful, and to a degree self-governing, but inconceivable without God as its designer and regulator.

THE LIFE SCIENCES

Western medicine experienced the same post-classical decline as the physical sciences – the closure of the schools in Rome and the break with Greek learning. The function of healing passed to the monasteries, but the Christian view of the body and its health tended to reverse the whole secular approach of classical Hip-

pocratic medicine, and the belief that disease was a divine punishment for sin was scarcely encouraging to scientific medicine. This was intensified by the popular cult of saints and their miraculous cures, where prayers were offered to certain saints for particular diseases – St Vitus for chorea (St Vitus's dance), St Anthony for erisypelas (St Anthony's fire), or St Roch for plague. The twelfth century saw a revival of Greek medicine, again with Arabic as the intermediary, and as well as Galen's, the works of Avicenna, al-Razi and Abu'l Qasim became the canonical authorities in the west. The first great western medical school in Salerno was traditionally and symbolically founded by four physicians, one Greek, one Arab, one Hebrew and one Latin, and was followed by Bologna, Padua, Paris and Oxford. As these institutions emerged into universities, the study of medicine took its place alongside that of logic, philosophy, theology and astronomy, and took shape within the medieval world-view, which embraced God, nature and man. It began its long transformation from a popular craft to a profession with a firm intellectual basis.

The central theory of sickness and health remained the Galenic one of the 'complexion' – the balance of the four humours, which were of course the four elements as manifest in the human body, cold or heat, moistness or dryness. The identification of any imbalance within the individual patient was detected by uroscopy, pulse and temperature iregularities, rashes and so on, and was treated with drugs, diet or blood-letting. Herbal medicines were predominant, but Dioscorides's works proved too detailed, and mentioned too many plants unobtainable in western Europe, therefore simpler herbals took his place. The great panacea was theriac, the substance brewed from vipers' flesh, originally as a cure for poison, and to which more and more ingredients came to be added, and some of whose popularity may have been caused by its opium content.

One of the primary causes of imbalance of the humours was considered to be celestial, and astrology came to play a central role in medieval diagnosis. The individual complexion was determined by the conjunction of stars and planets at birth, and this influence continued in the flow of celestial forces throughout life.

An anatomy lesson, illustrating a fifteenth-century copy of Avicenna's *Canon*. The bodies of executed criminals in particular provided medieval medical schools with the opportunity for dissection.

Glasgow University Library

Moreover the zodiacal signs were thought to dominate the different parts of the body, a scheme graphically expressed in the charts of the zodiac man, where Aries is assigned to the head, Leo to the heart, Virgo to the abdomen and so on. With this framework, the physician took account of the configuration of the heavens when the disease struck, when treatment, especially surgery, should be attempted, or when a crisis was approaching. Chaucer's doctor of physic could:

> '… guess the ascending of the star
> Wherein his patient's fortunes settled were.
> He knew the course of every malady,
> Were it of cold or heat or moist or dry.'

The influence of the heavens was even more likely to be detected in times of general diesease such as the Black Death of 1347–51, and the belief in the *physical* reality of this influence is seen in the judgement of the Paris medical faculty that the plague was caused by a 'corruption of the air' resulting from a conjunction of Jupiter, Saturn and Mars. Boccaccio too considered the plague to have been caused 'either because of the influence of the planets, or sent from God as a just punishment for our sins.'

Surgery was seen as on a lower, less intellectual level that this form of diagnostic medicine, and the anatomical knowledge of the middle ages showed no advance

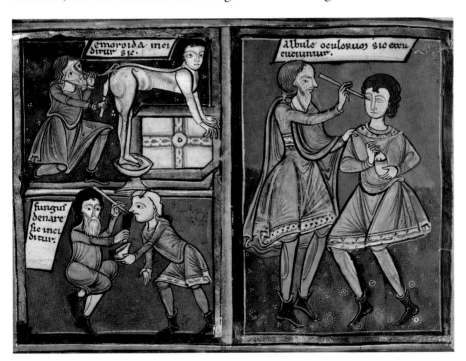

Medieval surgeons
operating for
haemorrhoids, nasal
polyps, and cataract.
The British Library
Sloane Ms. 1975 f.93r.

on the Galenic models. Dissection of the human body was carried out in the universities by the fourteenth century, usually on the bodies of executed criminals, but it was performed by an assistant, while the teacher demonstrated from texts, and the students merely watched: it was pedagogy and not research. The first Latin anatomical treatise, Mondino de Liucci's *Anatomia* of 1316, perpetuated certain Galenic errors, and even added new ones, but nevertheless contributed to a revival of first-hand anatomical studies, first at Bologna where Mondino taught, then more widely. Mondino's counterpart in the field of surgery was Guy de Chauliac, physician to several Avignon popes, and author of *Chirurgia Magna*,

1363, which is largely composed of quotations from Galen, al-Razi, Abu'l Qasim, and Avicenna. Guy described the plague at Avignon, the symptoms of the buboes in the armpits, and how it had spread from Asia, but, unable to transcend the spirit of his time, he ascribed the plague either to conjunctions of the planets or to a conspiracy of the Jews to poison the world. The data and the theoretical basis of medieval medicine were almost wholly Galenic and Arabic. Its original contribution lay in the elaboration of astrological theory, and the chart of the zodiac man was found deeply satisfying as a philosophical demonstration of the human body's place within the cosmos.

The sciences of natural history – botany and zoology – found their typical medieval expression in two textual traditions, the herbal and the bestiary. Both were securely based in observation of the real world, but both were overlaid with popular or legendary beliefs. These texts were illustrated with pictures which fixed the plant or animal in the reader's mind, and the artists almost invariably copied, and sometimes elaborated, existing models. Herbals were generally more accurate than bestiaries, for the plant is an easier object to capture, study and depict. Their primary applications lay in medicine and food, and although Dioscorides was known, a shorter and more practical work, the *Herbarius* of Pseudo-Apuleius, was the most common authority in the middle ages. The form of the plant was not described – that was left to the picture – but its situation and uses were, and particular interest was shown in any legendary or magical properties, the classic example being the mandrake, with its reputed scream when uprooted. The bestiary had perhaps a greater legendary component, for it included not only real creatures like the horse, fox and hedgehog, but the unicorn, the sphinx and the griffin too. The text of the bestiary also had a moral element, for the characteristics of the animals were described and contrasted in the manner of Aesop's fables, so that the fox is cunning to the point of devilry, the cranes operate in military formations, the lion can raise its stillborn offspring into life, and so on. The bestiary clearly had a function wider than that of a scientific treatise on animals: it was a repository of popular belief and legend, and it also served to interpret a world in which animals represented certain qualities from which man might learn. There are isolated examples of a more direct, empirical approach to the natural world, the most intriguing being the treatise on falconry written by the Hohenstaufen Emperor Frederick II, *De arte venandi cum avibus* – the 'Art of Hunting with Birds'. Hunting was an important activity in all courts and wealthy households, and the animals kept for the purpose were undoubtedly the means to a close familiarity with the animal kingdom, but one which rarely found its way into books. The text deals with the migration, anatomy, nesting, flight and training of birds, all drawn from keen personal observation, for this was a field where the great names of Aristotle or Avicenna for once had nothing to say.

The basilisca (adderwort or snakeweed) from a twelfth-century manuscript of the *Herbarius* of Pseudo-Apuleius. The plant was reputed to grow where the basilisk snake resided, and its roots were said to resemble a claw-foot.
The British Library
Sloane Ms. 1975 f.48r.

SCIENCE IN A RELIGIOUS CULTURE

The outstanding fact about the culture of the middle ages in the west and in the east was the centrality of religion. The terms Christendom or Islam defined two great empires of the mind and spirit, while the Bible and the Koran, the cathedral and the mosque, the crusade and the *jihad* were the dominant features of their landscapes. The scientific thought of these cultures was strongly marked by a quest for universals, for some order of conceptual reality which lay behind that which was seen in the natural world, which made it coherent in itself, and comprehensible to the human mind. Such universals included circular motion or the elements of the material world or the form-substance duality. Such a quest has in a sense a very modern ring, for is not twentieth-century physics seeking to uncover the forces which bind the material universe together? The difference is that medieval science did not approach these laws from an empirical basis, for they had virtually no analytical data to build on. Instead they began with the conviction that the universe is the creation of a rational mind, and that it must be organized according to rational principles. The principles thus formulated were those which satisfied minds accustomed to metaphysics, accustomed that is to thinking of a world permeated by real but intangible forces. The universals were discovered by logic, by inferences held to be self-evident, which were never tested in the natural world, for the means for such testing did not exist. Medieval thinkers constructed a coherent system of thought about a universe which was vast but finite, complex but purposeful, and one which was, partly through intellectual discipline and partly through the assent of faith, comprehensible to man. This is the vision embodied in Dante's *Divine Comedy,* especially in the last part, the *Paradiso,* where, free of earthly constraints and guided by real figures such as Aquinas, Dante journeys through the heavenly spheres, awed by the majesty of the cosmic design.

Such a cosmology, however ingeniously worked out by Ptolemy and his successors, was purely descriptive and external: there was not, and could not be, any real penetration into the forces which sustained it. This was inevitable, for the terms in which questions were posed and answered by the medieval scientist were not those we would use. The material culture and the technology of the middle ages will illustrate this: there was simply no cause to think in terms of mass or velocity or mechanical energy or fields of force or gravity. It is often said that medieval philosophers neglected experience, but it would be truer to say that their experience was fundamentally different from ours. It is striking how any discussion of medieval science centres on texts – on the transmission and absorption of the ideas of others, ideas which could be subject only to logical scrutiny, not to quantitative testing. How much in western medieval science is *not* traceable to Greek or Arab sources? The medieval respect for authority meant that novelty of information or ideas was less highly valued than mastery of traditional sources. The leading philosophers of medieval Europe spent their entire careers making elaborate commentaries on thinkers who lived and died fifteen centuries before. This is surely related to the fact that the medieval world, its material culture and visible landscape, would have been entirely recognizable to the ancient Greeks. Change, progress, originality, innovation, revolution – these were not aims which would have been acknowledged to outweigh the perpetuation of inherited learning. In search of a paradigm of medieval science, one need look no further than the *mappa mundi,* the world map where real geography appears side by side with images from legend and from the Bible. Fragments of reality are there – countries, seas, cities – but their vital relationships are unmeasured and amorphous. If a scientific renaissance was to take place, as we know it did, it would emerge only slowly from such roots, just as those medieval inventions, the compass, the clock and the gun, grew slowly from novelties into tools capable of re-shaping the material world.

Chapter Four

THE PROBLEM OF THE RENAISSANCE

&❧

'I count religion but a childish toy,
And hold there is no sin but ignorance.'

Marlowe, *The Jew of Malta*

PAINTERS AND MYSTICS

The Renaissance seems at times to be a game anyone can play: historians of art, of philosophy, of economics, of science all enjoy refining its definition or denying its very existence. Contemporary observers, especially Italian writers from Petrarch to Vasari, had no doubt that they were living in an age of rebirth; but what precisely did they mean by that word? Was this rebirth a purely artistic event, or did it penetrate more deeply into the thought and spirit of the age?

In Italian paintings of the 1460s and 1470s, the visual symbols of the Renaissance are now easy to recognise: the perspective of the courtyard settings, the animals and flowers drawn with fidelity and vigour, and above all the gracefully-proportioned human forms: a new engagement with the natural world is powerfully evident. But was there, behind these images, any new scientific knowledge of geometry, biology or anatomy? The answer is almost entirely negative. Any trace of a renaissance of science paralleling that of the arts is extremely difficult to demonstrate. The artists appear to have made their own rediscovery of nature long before scientists began to look with new eyes at the structures of the physical world. Vasari defined Renaissance art as 'the intelligent investigation and zealous imitation of the true properties of the natural world.' This sounds like a charter for science, yet Leonardo, Michelangelo and Dürer had wrought their revolution in art and had passed away years before the publication in 1543 of Copernicus's *De Revolutionibus*, the manifesto which inaugurated a renaissance in scientific thought. And that same publication came a full century after the greatest single technological innovation in recorded history, the advent of printing. So the central problem of the Renaissance in scientific terms is this: how is it that the movement of European rebirth, which is always believed to stand at the opening of the modern era in world history, had apparently no element of the characteristic force which has shaped that era, namely scientific knowledge?

It is clear that by the eventful year of 1543 (also the year of Vesalius's great work on anatomy) the context of scientific thought had already been transformed by a complex of events, which we call the Renaissance. What were these events? First, urban communities, especially in Italy and south Germany, were experiencing new levels of wealth, which in turn fostered new architecture, new scholarship, new social networks and new intellectual challenges. Second, artists had rediscovered both the natural world and the means to depict it through their mastery of perspective, and their works inspired contemporaries to look with new eyes at nature's forms and forces. Third, European navigators had revolutionized man's picture of his world, revealing oceans and continents that had lain hidden throughout history. Fourth, the new technology of printing made possible the interchange of ideas and information on a scale previously undreamed of. And fifth, the Protestant Reformation had shattered the unity of medieval Christendom. While it is difficult to chart the way these factors related to each other, they evidently combined to create in the sixteenth century a climate of intellectual *innovation*, which was in complete contrast to the medieval respect for authority and continuity. Perhaps the most difficult factor to assess is the Reformation, for despite its evident challenge to the authority of centuries, no simple correlation exists between Protestantism and intellectual freedom, scientific or otherwise. Moreover, had the Reformation not awakened the Catholic church into a campaign against heresy in all its forms, it is possible that the church might not have been moved to oppose the theories of Copernicus and Galileo. It seems generally true that a new approach to nature was emerging in the fifteenth century, provoked by the social and technical changes listed above. The innovative science of the period was predominantly in what we should call the applied sciences – in fields such as medicine, navigation, architecture, mining and engineering – where new observations and new techniques preceded any intellectual revolution in pure science.

One of the major components in all these fields was the emphasis on mathematics as a professional tool. The prosperity of Renaissance Italy, the spread of credit and banking, and the existence of several regional systems of coinage, had all created a demand for mathematical skills, facilitated by the Arabic numeral system. The first printed text of Euclid appeared in 1482, and numerous works of practical arithmetic were published in the following decades, such as those of Luca Pacioli, who wrote on book-keeping, money, weights and measures, as well as algebra and geometry. Pacioli was in contact with the artists Piero della Francesca and Leonardo da Vinci, indeed Leonardo drew the geometric figures for Pacioli's most famous work *Divina proportione* ('On Divine Proportion', published 1509). The concept of proportion as inherent in nature and capable of being discovered and imitated by artists was fundamental to the Renaissance approach to painting and architecture. Pure mathematical research in Italy was furthered by Niccolo Tartaglia (1499–1557) and Girolamo Cardano (1501–1576), who pioneered cubic equations, although algebraic notation as we now know it was still not available, and the workings were presented verbally, in a series of logical steps. Cardano also wrote encyclopedic works on technology and natural science, works which bear a close resemblance to some of the ideas of Leonardo da Vinci, and which have led historians to suppose than Cardano may have had access to Leonardo's unpublished notebooks. Both Cardano and Tartaglia were concerned with the new science of ballistics, and the questions about Aristotelian physics which it raised: was the force which maintained a projectile in motion really the movement of the air around it, and was its path simply rectilinear – a straight line of ascent followed by one of descent? Both men came to the conclusion that the path of a cannon-ball is a curved line throughout its length, but they did not go on to draw theoretical conclusions about force, acceleration or gravity which might have anticipated Galileo. By the mid-sixteenth century however, it is clear that mathematics was increasingly being seen as a means of expressing concepts such as proportion, measure and movement.

Two systems of arithmetic: the man on the left is calculating with the new Hindu-Arabic numerals, and is clearly working faster than the disconsolate figure on the right, who is still using the traditional abacus. From Reisch's *Margarita Philosophica* 1503.

The British Library 54.c.14.

Aiming canon using quadrants, 1547. The mathematician Tartaglia had argued correctly that the maximum range would be attained with a firing angle of 45°. The study of ballistics played an important part in Renaissance thinking on matters such as force and gravity. From Ryff's *Mathematischen und Mechanischen Kunst*, 1547.
The British Library 1262.g.21.

There was one form of natural philosophy – Hermeticism – which has often been considered peculiarly the creation of the Renaissance, and lauded as an anticipation of modern science. In fact Hermeticism was more of a gnostic, semi-mystical system than a science, and it has appealed to the modern imagination more by virtue of its audacity than its scientific rigour, while a powerful legend has gathered about its leading figure, Giordano Bruno. Hermeticism took its name from a body of teachings supposedly written by the Egyptian deity Thoth, whom the Greeks identified with Hermes. These Greek texts were known to the early church fathers such as St Augustine, and were believed to be of immense antiquity, although they were later proved to to date from the second century AD. They dealt with the creation of the world, with astrology and medicine, and with alchemy, magic and ritual. Certain passages referring to the 'Son of God' were taken to be prophecies of Christianity. These Hermetic writings were known to few in the west until the Platonist, Marsilio Ficino, translated them into Latin in 1471, and created a surge of interest in them which would last for almost two centuries. Their central teaching was that man and the cosmos were interdependent, and that laws of sympathy or affinity existed which linked man to both the natural and the spiritual world. In this system, knowledge of nature's secrets would lead to the release of the soul, to the ennoblement and almost the deification, of man. To the extent that this belief stimulated the practices of alchemy and astrology, there are some slight grounds for regarding Hermeticism as a form of proto-science. But this is outweighed by the fact that these assumed secrets and affinities could never be discovered by ordinary scientific methods, hence magic and divine revelation were invoked. Hermeticism was in fact a type of gnostic religion, with the emphasis on secret, arcane knowledge, available only to a devoted elite, and as such it tended to foster the pursuit of magic rather than secular, objective science. This aspect of Renaissance science is consummately expressed in the legend of Faust, who, having exhausted conventional scholarship,

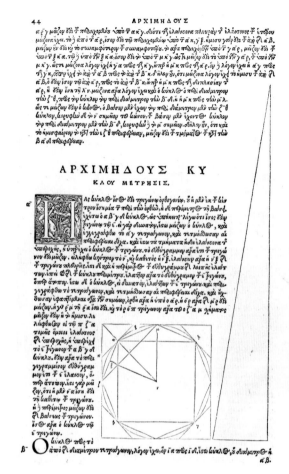

Left: The first printed edition of Euclid issued in Venice in 1482. The text used was the Latin translation by Adelard of Bath, revised by Campanus of Novara in the thirteenth century.
The British Library IB 20513.

Right: The first printed Greek text of Archimedes, 1544. The rediscovery of Archimedes was influential in turning the attention of Renaissance scientists such as Galileo to the mathematical analysis of physical problems.
The British Library C.82.h.9.

turns fatally to magic in order to penetrate nature's secrets. It has been plausibly argued that Hermeticism's heady vision of man's mastery over the forces of nature anticipates not modern science but modern technology, whose effect is to release us from servitude to our physical environment.

The modern interest in Hermeticism probably arises from a single prophetic idea – the infinity of the universe – associated with the most famous of the Renaissance Hermetic philosophers, Giordano Bruno (1548–1600). Bruno was a turbulent figure, first a Dominican then a Calvinist then a gnostic, roaming Europe teaching and writing, and making intellectual enemies everywhere – in Italy, France, England and Germany. Turning his back on orthodox religion, Bruno wrote in Neoplatonic terms about the soul's attainment of unity with the infinite. The religion of the Bible, he argued, existed merely to govern ignorant people, while philosophy was the path of the elite. He revived the Platonic concept that the cosmos was a living organism, animated by a soul. His most original venture into scientific theory was to suggest that the universe was infinite, and contained a multiplicity of worlds. Bruno based this idea not on any astronomical evidence, but on a mystical intuition that the world-soul could not be contained in the puny mechanism of spherical astronomy. Bruno was arrested by the Inquisition, and after eight years of imprisonment and interrogation, was burned in Rome as a heretic in 1600. The precise charges against him we do not know, for some of the principal documents are lost, but it is unlikely that his cosmology was

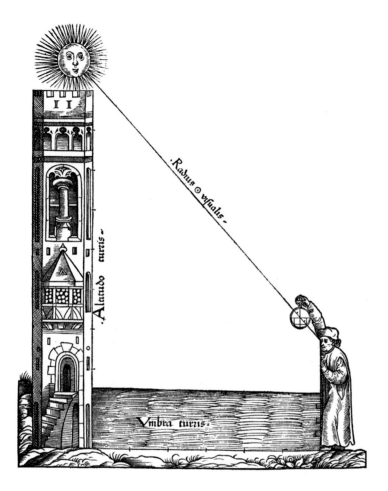

the cause of his death: he had often implicitly denied the divinity of Christ, and was also suspected of diabolical practices. The image of Bruno as a martyr for science, refusing, unlike Galileo, to retract what he knew to be scientifically true, is a false one. Bruno stood quite outside the tradition of mathematical astronomy, and the view that this Renaissance mystic was a forerunner of modern science is untenable, except in the sense that his philosophy of nature challenged the ruling Christian one.

A modern fascination, less powerful than that of Bruno but still considerable, has also gathered around the life and thought of Nicholas of Cusa (*c.*1401–1464), and, coincidentally, on similar grounds. Cusa, a canon lawyer, historical scholar and cardinal of the church, evolved some logical and mathematical theories that have a very speculative, medieval appearance, such as the 'coincidence of opposites'. This principle argued, for example, that a straight line and a circle may become identical at infinity, since the circumference of an infinitely large circle will be a straight line. On this same principle, the point generates lines, surfaces and solid forms, therefore the smallest point may be said to mirror or contain the largest form, even the whole universe. This metaphysical

Measuring the height of a tower using an astrolabe. The astrolabe is used to construct a triangle similar to that formed by the tower and its shadow. From Stöffler's *Elucidatio .. astrolabii,* 1512.
The British Library C.54.f.25.

reasoning would be of no more than academic interest were it not for the fact that Cusa drew the intriguing conclusion that the universe was infinite, that the earth was not its central point, and that the movements of the cosmic bodies were not absolute, but were relative to the observer. It was this theory, a mere curiosity in its own time, which led much later to Cusa's being seen as a herald of Copernicanism. Yet Cusa's theory was not supported by any astronomical evidence, and remained only a speculation, a logical possibility. In fact Cusa emphasized the lack of mathematical exactness in ordinary experience – the absence of truly straight lines or perfect circles in nature – and accordingly he argued that a mathematical description of nature was impossible. Cusa the prophetic scientist vanishes, to be replaced by the subtle logician, weaving a private intellectual tapestry of his own. By contrast, in the year of Cusa's death, certain advances were under way in true mathematical astronomy which can be justifiably seen as the first, and rather isolated, stirrings of a renaissance in pure science.

THE REFORM OF ASTRONOMY

Astronomy in the mid-fifteenth century involved very little empirical, observational work, and its main function was to underpin the practice of astrology. The astrologers' work consisted of calculating the positions of celestial bodies in order to interpret the human events which they believed were governed by those positions. Optical sighting devices for plotting celestial positions did exist, but most astronomers relied on pre-calculated tables to give them the data they needed. The Ptolemaic theory of cycles and epicycles, with the assumption that all bodies moved in uniform circular motion, had enabled scholars to calculate all the positions which the astrologer needed. The most widely-used of these tables were the Alfonsine Tables, compiled in Spain for the epoch 1252. Thus the astronomer-astrologer of the fifteenth century was distanced from the direct study of the heavens by the rigid, unquestioned theory of cosmic structure, and by pre-computed data.

One of the major inspirations of Renaissance art was the re-examination of classical models, and this process was now repeated in the field of astronomy. Ptolemy's great astronomical text, the *Almagest,* was far from unknown in medieval Europe, indeed the cosmic structure which it describes underlay all the cosmological thought of the age. But the *Almagest* is a long, technical and highly demanding work, which few individuals in Europe before the late fifteenth century were capable of mastering or challenging. This situation changed in the years 1455–1465 with the innovative work of two German astronomers, Georg Peurbach and Johannes Müller, known as Regiomontanus (the Latin form of his home town Königsberg). Working in Vienna, they undertook a comprehensive revision of the *Almagest,* based on a number of careful observations of planetary positions, and a thorough re-working of its mathematics. They then compared their findings with the tables of astronomical positions used almost universally in this period, the Alfonsine Tables. Peurbach died when the task was less than half finished, but Regiomontanus carried it to completion by 1463. The resulting *Epytoma in Almagestum Ptolemei,* appeared in 1496, and its impact as the first printed work of Ptolemaic astronomy was enormous. This work was a vital one in that it revealed numerous errors in the Alfonsine Tables, errors so serious that they provoked Regiomontanus to exclaim 'The common astronomers of our age are like credulous women, receiving as something divine and immutable whatever they come upon in books … for they believe in writers, and make no effort themselves to find the truth.' The motions of all the planets were found to be incorrectly plotted in the Tables, but Regiomontanus attributed no blame to Ptolemy; rather he considered that the reform of astronomy would be achieved by applying Ptolemy's mathematical models to new, more precise observation. Regiomontanus set in train a process whose end would have astounded him, for he revealed discrepancies between the astronomical authorities and observed reality which would be solved only after the revolution wrought by Copernicus, Tycho and Kepler. Copernicus himself studied the *Epytoma,* and was struck by these discrepancies: Copernicus however concluded that it was the Ptolemaic framework itself which was faulty.

Regiomontanus settled in Nuremberg and established a printing press with which he planned to disseminate scientific texts, but he died prematurely in 1476, probably a victim of the plague, in Rome, where he had gone at the request of Pope Sixtus IV to assist in the reform of the calendar. Two years earlier he had published his *Ephemerides,* tables of celestial positions for the years 1475–1506, which established themselves at once as the most accurate in existence, and were known to been used by navigators, Columbus among them. Regiomontanus was a transitional figure, standing on the threshold of a scientific renaissance. He did not dream of questioning Ptolemy's cosmic model, or the geometric detail with which it had been worked out, but his original observations and the calculations he made from them opened the way for his successors' radical re-appraisal of classical astronomy. He introduced an important new note of empiricism into a science which had become enslaved to texts and tables. By his observations and by his mastery of mathematics, he was able to show that the astronomy of his day was built on sand. Had he lived, the steady accumulation of data might have led him to anticipate Copernicus. He founded in Nuremberg a school of mathematical science which included figures such as Martin Behaim, Johann Schöner and Bernhard Walther, who continued the original publishing programme, culminating in the first printing of the Latin *Almagest* in 1515.

Regiomontanus's plans to use the revolutionary medium of printing to disseminate a fresh basis for astronomical research justify our placing him at the head of a new phase of scientific development. It is a paradox that of the invention of printing itself, with its

Ptolemy (left) and Regiomontanus beneath a giant armillary, the model of the cosmic structure in classical and medieval thought. From Regiomontanus's *Epytoma in Almagestum*, 1496.
The British Library IB. 23380

almost incalculable importance for intellectual history, we know so little. It is often said that printing was practised in China for centuries before its arrival in the west, and indeed there was a long Chinese tradition of replicating both written characters and images from incised wooden blocks. But the nature of European printing was radically different, consisting of the flexible use and re-use of hundreds of textual elements, this being dictated by the nature of the Roman alphabet, so different from the Chinese characters. It was Johannes Gutenberg (*c.*1390–1468) of Mainz who conceived this method of fragmentation, and who worked in close secrecy for some fifteen years from 1438 onwards, to overcome the enormous technical difficulties. His eventual success lay in four vital developments: the accurate casting of type in a soft alloy from a very hard matrix; the technique of locking-up the type in a strong frame; the modification of the crude wine press to give the necessary spread of pressure; and a fluid yet stable form of ink. None of these elements existed in China or in Europe before Gutenberg, and by 1455 he issued his monumental Bible, the first printed book. Although no scientific theory was involved (as far as we can tell) in Gutenberg's work, his achievement is of immense interest as being the first individual invention in the modern mould: where a new and complex system is conceived, designed and built with the deliberate aim of changing – revolutionizing – an existing practise. With his originality of conception and his long years of determined experimentation, Gutenberg stands as an unprecedented figure.

THE ARTIST AS SCIENTIST

It is a striking fact that for the first half-century of its existence, the new medium of printing took for its subject not new literature and new ideas, but classical and religious texts. To philosophers, artists or scientists engaged in original work, the printing of their results was not yet a natural event, and their ideas remained long in manuscript. Nowhere is this more clearly evident than in the notes, drawings and dreams of the archetypal Renaissance figure, Leonardo da Vinci (1452–1519). Leonardo himself published no scientific works, but his extensive manuscripts, almost entirely unknown until the nineteenth century, uniquely preserve all the sketches, speculations, errors and false starts, normally destroyed when a finished work reaches printed form. In a sense Leonardo has no formal place in the history of science because his ideas remained unknown for centuries; yet his private studies of anatomy, biology, mechanics, geology and mathematics make it clear that his overwhelming concern was a search for the forms and structure, the geometric rules, which he believed to underlie all nature. The apparent disarray of his notebooks sprang from his characteristic method of study, for on any subject it was his habit to amass diverse examples and analogies which would reveal the underlying patterns, the hidden logic of natural forms; thus a page devoted to curves might include sketches and notes on waves, grasses, hair, screw-threads and sails. Leonardo repeatedly stated that the artist's eye should serve this quest for understanding: 'painters study such things as pertain to the true understanding of all the forms of nature's works, and earnestly contrive to acquire an understanding of all these forms'; 'the eye, the window of the soul, is the chief means whereby the understanding can most fully appreciate the infinite works of nature'; 'painting compels the mind of the painter to transform itself into the very mind of nature.' Mathematics, and especially geometry he considered to be at the heart of these forms, to be the hidden language of nature. In his analysis of vision, Leonardo suggested that light was reflected from any object in circles which spread like ripples in still water: 'just as a stone flung into water becomes the centre and cause of many circles … so any object placed in the luminous atmosphere diffuses itself in circles and fills the surrounding air with images of itself'. His work on vision led him to dissect the brain, and he was perhaps the first to attempt to map it for psychological functions such as memory and judgement.

It is Leonardo's attention to anatomy that has left its clearest mark on his art, which shows an unprecedented mastery of the human form. Yet his devotion to truth and nature cost him much anguish in this respect, and he records 'passing the night hours in the company of corpses, quartered and flayed, horrible to behold.' His dissections led him to theorize extensively on physiology, and he made models of the heart out of wax and glass in an unsuccessful attempt to

Artists studying and sketching at a school of anatomy *c*.1560 by Bartolomeo Passarotti. By this date, such study was deemed an essential part of an artist's training.
The Louvre, Paris

understand its workings. Equally famous are his investigations into the mystery of flight, his detailed studies of the bones and muscles of birds, and his attempts to arrive at mathematical formulae of wingspan and weight, which he might then transfer to man. Experiment was crucial for Leonardo, and in this context he placed a man with artificial wings on a huge set of scales in order to measure his decrease in weight when the wings were flapped downwards. He would use markers to observe forces in the movement of air or water. If experiment falsified an accepted rule, he would reject the rule, for example he disproved the accepted Aristotelian doctrine that 'if a power moves an object with a certain speed, it will move half that object with double the speed'. Although not a skilled mathematician – he was entirely self-taught in this field – he was convinced that 'there is no certainty where one cannot apply any of the mathematical sciences', and that 'proportion is found not only in numbers and measurements, but also in sounds, weights, times and spaces'. His notebooks contain many geometric representations of motion, weight and speed.

By profession Leonardo was known as an engineer, a surveyor and architect,

Dürer's perspective drawing of a lute, 1525. The artist touches one end of a string on various parts of the lute, and the point where the line intersects a window is noted. The joining of these points will give a precise, foreshortened view of the lute. The window represents the eye, and the string represents the individual rays of light impinging on it. From Dürer's *Unterweysung der Messung*.
The British Library C119.h.7(1)

more than as a painter, and his inventions in this field were precocious, although it many cases his ideas remained only at the design stage and were never actually built. He designed gears, bearings, lathes, presses, canal locks, textile-spinning machines, thread-cutters, weapons, and flying machines. He made a form of plastic by impregnating laminates of paper with gelatin and egg-white, producing hard, water-resistant, unbreakable plates. His inventive genius was sustained by a genuine understanding of mechanics: he had an intuitive grasp of the principle of the conservation of energy, and knew that the quest for a perpetual motion machine was an impossible dream, since machines do not initiate work, but simply modify its application. He apparently had little interest in the higher flights of cosmology – was this because it could not be subject to experiment? – but some of his most original theories were concerned with geology, where he elaborated a parallel between man and the earth: 'as man contains within himself bones, the supports and armature of the flesh, the world has rocks, the supports of the earth; as man has in him the lake of blood, which the lungs swell and decrease in breathing, the body of the earth has its oceanic sea, which also swells and diminishes every six hours in nourishing the world'. A great model-builder, he made a model of the Mediterranean in order to demonstrate the effect of the rivers flowing into it. Leonardo reasoned that the face of the earth had been formed by storm, flood and sedimentation over periods of time completely beyond the biblical time-scale, estimating for example that 200,000 years would have been necessary for the formation of the plain of the river Po. Through the richness of his thought, Leonardo is a figure of endless fascination, and one who presents the historian

with an insoluble problem: how can he be claimed as the essential representative figure of the Renaissance, when he was clearly so exceptional, an isolated genius working in private, without teachers, without pupils, and without measurable impact on the history of science? Yet he was after all a man of the Renaissance, for no Leonardo had been born in the fourteenth century or the thirteenth or the twelfth; somehow he must be seen as a prophetic figure, sharing a newly-emergent faith in observation and experiment, and in the form and logic of nature.

Less wide-ranging than Leonardo, his northern contemporary Albrecht Dürer, was equally convinced that form and proportion were the keys to nature's structures, and he brought his theories to expression in several published works on mathematics, on human form and on architecture. Like Leonardo, Dürer at first studied nature in order to be able to render its forms accurately, but that study led him to conclude that mathematical rules of proportion were built into nature. Both Dürer and Leonardo were indebted to the ealier theorist of Renaissance art, Leone Alberti, whose text *On Painting* of 1435 expounded the theory of perspective that was so central to the artist's imitation of nature. Alberti echoed the view of the great sculptor Ghiberti, that the artist who wished to render the human form correctly must 'have witnessed dissection in order to know how many bones are in the human body, what are the muscles in the body, and similarly the tendons and sinews'. After expounding his theory that the human figure provides the essential frame of reference for the scale of all other forms, Alberti goes on to urge that the techniques of sculpture, architecture, perspective and engineering be focused into the harmonious ordering of urban space, the Renaissance quest for the ideal city.

SCIENCE FROM THE EARTH

Experiment and theory meet in the work of one of the most singular and perplexing of Renaissance thinkers, the physician and mystic, Paracelsus (1493–1541), although theory is too prim a term for the opulent mysteries of his mind. Born Philippus von Hohenheim, his family background and early training lay in Alpine mining and geology. He studied medicine at German and Italian universities, adopting the name Paracelsus to indicate that he had transcended Celsus, the Roman medical author. As a military surgeon he travelled far, to Scandinavia, Russia and the Middle East, before settling to practise in Basel. Paracelsus resembled Giordano Bruno in his turbulent quarrels with authority, and is reported to have publicly burned the works of Avicenna, an act which gave him the reputation of being a scientific Luther, recalling the reformer's burning of the Papal Bull. Forced to leave Basel, he never again obtained an academic or any other settled position, but achieved fame with his book *Die Grosse Wundartzney* ('The Great Treatise on Surgery') in 1536. He pioneered the tretament of syphilis with measured doses of mercury; he demonstrated that the miners' disease, silicosis, resulted from inhaling dust particles; and he diagnosed the problem of goitre as the composition of drinking-water in mountain regions.

Central to Paracelsus's intellectual rebellion was his rejection of the classical notion of the humours, in which all disease was traceable to an imbalance of the four elements of heat, cold, moisture and dryness in the body. Each individual had his own unique balance, or temperament, and there were therefore as many diseases as there were individuals. In this scheme, diseases were not classifiable as objective, functional disorders with their own specific agents. Paracelsus rejected this understanding, and argued that diseases did indeed arise from direct causes in

The Paracelsian cosmos, in Johann Mylius, *Opus Medico-Chymii,* 1618. Animals, plants, minerals and heavenly bodies are linked in a system of correspondences – note the chain by which the man and woman are bound to the heavenly sphere.
The British Library 1033.l.4, 3.

Portrait of Paracelsus.
Science Museum, London

the external world, that disease was a parasitical agent which invaded the body and threatened its life. This sounds remarkably modern, and prophetic of our understanding of disease; but what, in Paracelsus's view, were these agents? Having traced the progress of disease in natural terms, Paracelsus proceeds to give them a supernatural explanation: they were poisons of various kinds which emanated from the earth and the air and the stars, as astral bodies, using the elements of this earth as vehicles for their actions. Chemicals derived from plants and minerals can be used to combat these agents, for example mercury was known to cure syphilis by expelling poison salts from the body; but in reality it is the

An alchemist, from Ashmole's *Theatrum chemicum Britannicum*, 1652. The balance has a very scientific appearance, but the symbols of the Sun and Moon upon the desk indicate that this practitioner is seeking to isolate the mystic properties of the noble metals.
The British Library E.653.

celestial virtue of Mercury that is operating. This was a form of astrology, but one in which astral forces work through specific and identifiable physical channels. Paracelsus believed that all nature possessed its own 'knowledge' – the knowledge of the seed how to grow into a tree, the knowledge of minerals which grow in the earth – in which man can learn to participate. He conceived an invisible world parallel to this one, teeming with spirits, which operated in natural, analysable ways. Thus the whole concept of nature is broadened into a pantheistic religion of a living cosmos. The sources of this complex, spiritualised theory of nature lay undoubtedly in the neoplatonic and gnostic literature of the Renaissance, to which Paracelsus added an original medical dimension. He was an alchemist, but not in the familiar sense of one who attempts to transmute metals, but as seeking

new chemical methods to combat disease. He developed the alchemical divisions of matter into 'principles', primarily mercury, sulphur and salt, which were both physical substances and occult forces representing the struggle between spirit, soul and body. Part of the legacy of Paracelsus, especially the concept of the astral body, entered the European occult tradition, and his reputation became a deeply ambiguous one. But the remarkable aspect of his career was that he combined the exploration of this mystic labyrinth with a life of constant experiment, and became one of the founders of chemistry. In rejecting the theory of the humours, he stimulated others to consider new models of disease, and his influence is traceable in scientists and physicians through the seventeenth and eighteenth centuries, in the work of Van Helmont, Boyle, Stahl and Boerhaave. The age of Paracelsus was the high noon of alchemy, when philosophers all over Europe attempted to prise open nature's secrets with the mortar and pestle, furnace and distillation jar. Yet their efforts were vitiated by their enslavement to abstract or occult principles which had no basis in fact, and which led them to cloak their theories in an elaborate symbolic language, comprehensible only to the initiated, and utterly removed from the language of science.

Exactly contemporary with Paracelsus was another German physician, Georgius Agricola, whose work in geology and mining resulted in one of the seminal texts in the history of science and technology. Agricola (1494–1555) studied medicine at Italian universities before returning to his native region of the Erzgebirge in Saxony-Bohemia, a centre of iron, copper and silver mining. Here

Assaying laboratory, from Agricola's *De re metallica*, 1556. Agricola's comprehensive descriptions were accompanied by scores of such detailed pictures, making it an invaluable record of Renaissance technology.
The British Library C.81.h.11.

his interests broadened from the many medical ailments of those who laboured in the underground chambers and the smelting-houses, to the earth itself, and for twenty-five years he gathered materials for a series of books, culminating in the posthumous *De re metallica* (1556), the definitive statement of sixteenth-century knowledge of geology, metallurgy and mine-engineering. Agricola's approach was avowedly empirical: 'Those things which we see with our eyes and understand by means of our senses are more clearly to be demonstrated than if learned by means of reasoning.' He rejected many traditional fables about the substance of the earth: the biblical doctrine that the earth had been formed complete at one point in time; the medieval theory that metals grew from seeds, like plants; the astrological notion that minerals were attuned to the stars and planets; and the alchemist's belief that the base metals were transmuted into the noble ones, gold and silver, within the bowels of the earth. Agricola could offer no comprehensive account of geological change, but he reasoned correctly that mineral veins were the remains of aqueous solutions, that rocks and minerals were metamorphic, that is, new forms emerged as old ones decayed, and that heat and water were the primary agents of change. One of his aims was to identify the hundreds of substances known by different names in the different traditions – classical, medical, alchemical and mining – which he did by tests such as colour, smell, and hardness. In Agricola's time there were only six known metals – gold, silver, iron, copper, tin and lead – but he argued that mercury should be added, and he thought it certain that many other unknown metals existed. Agricola gave full descriptions of extraction, purification and assaying, and one of the distinctive features of his book was the many detailed woodcuts of these techniques and the equipment required, for as he said ' I have hired illustrators to delineate their forms, lest descriptions which are conveyed by words should either not be understood by men of our own times, or should cause difficulty to posterity, in the same way as difficulty is caused to us by the many names which the Ancients have handed down without explanation.' Agricola witessed a considerable revolution in the use of water power in mines, and his book is full of images of ingenious early pumps, winches and power trains, and perhaps most notably, of the earliest type of railroad, in which trucks were retained on a wooden track by means of grooves and guides; the trucks were of course pulled or gravity-driven. It was impossible for Agricola to develop single-handedly a geological theory to underpin his description of metallurgy, but his empirical approach to the mysteries of the earth was a landmark of Renaissance science. This seminal work is complemented by a related text of 1540, *De la pirotechnia* by Biringuccio, which described smelting, metal-casting, and the making of gunpowder and fireworks. Like Agricola, Biringuccio inhabits the world of the alchemist, the world of chemicals and furnaces, but they have both stripped away the dreams and the occult jargon of the alchemist, and are concerned solely to give a clear, factual account of their procedures.

THE BIOLOGICAL SCIENCES

An impulse towards more objective and more complete classification is also evident among biologists of the mid-sixteenth century, and a series of works began to appear from 1530 onwards whose descriptions of plants and animals at last went beyond the medieval herbals and bestiaries. There are several possible reasons which suggest themselves for this. First, that the discoveries in the New World had greatly enlarged man's view of the varieties of flora and fauna, after

Red poppy, *Papauer rubrum,* from Brunfels's *Herbarum,* 1530, one of the earliest printed works of illustrated botany, distinct from the herbal tradition.

The British Library 449.k.2.

Botanical artists at work from Fuchs's *De historia stirpum,* 1542. Fuchs's work was based mainly on established classical authorities, but its originality lay in its many vivid illustrations drawn from life. Albrecht Meyer (right) draws a corn-cockle, while Heinrich Füllmaurer transfers the drawing to a woodblock.

The British Library 36.h.8.

centuries in which European natural history had been essentially that of the ancient, classical world. Second, the artists of the Renaissance were producing resplendent images of the living world in comparison with which the pictures accompanying biological texts were vague and inaccurate. Third, the advent of printing created the ideal medium through which images of animals and plants might be disseminated and discussed. Priority among this group of biologists belongs to Otto Brunfels (1489–1534), whose three volumes of *Herbarum vivae eicones* ('Illustrations of Living Plants'), 1530–1536, contained woodcuts of 238 plants, drawn with a wholly new level of accuracy. Brunfels's text cannot be called original, being mainly derived from older authors, not does he attempt any form of systematic classification. The importance of the work undoubtedly rests on the fidelity of the pictures, by the artist Hans Weiditz, in which signs of the plants' wilting or being damaged are sometimes visible, leaving no doubt that they were drawn with the utmost care from living models. Brunfels's work was closely followed by that of Jerome Bock (1498–1554) whose *Neu Kreutterbuch,* was very different in kind. It was written in German and its first edition contained no pictures, but the text was based far more on personal observation, and it included many new species, where older authorities including Brunfels found in nature only what they had previously seen in Dioscorides and others. Bock attempted to group plants according to their form, and gave verbal descriptions precise enough to identify plants in the field. The criteria on which he based his taxonomy were the form of the plant, the shape of the corolla, and the formation of the seed capsules. Bock, like all his contemporaries, was unaware of the sexuality of plants, so that he could not achieve a valid taxonomy. The third of the 'German Fathers of Botany' was Leonhard Fuchs (1501–1566). Fuchs placed great reliance on classical authorities, and he was consequently a less original observer than Bock. His *De historia stirpium* ('The History of Plants', 1542) was even more richly illustrated than Brunfels's book, and contained many plants described for the first time (although the fuchsia, imported from the Americas in the seventeenth century, was named after him, and not by him).

PICTORES OPERIS,
Heinricus Füllmaurer.   Albertus Meyer.

The animal kingdom also had its Renaissance illustrators, although a wild animal is clearly less accessible and less passive than a plant as a subject of art, and certain misconceptions and legends tended to persist for longer, the classic example being the segmented armour-cladding of the rhinoceros, stemming from Dürer's

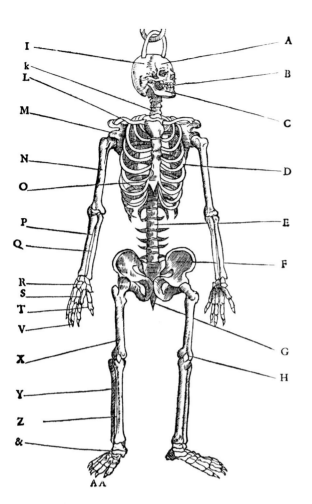
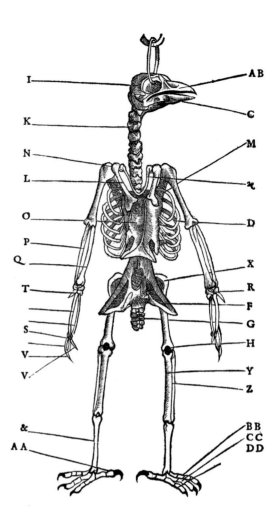

woodcut of 1515. Such pictures were often based on eye-witness sketches, which then were published and re-copied for years. Pierre Belon (1517–1564) published descriptive books on fish and on birds which contained many important original observations. He had dissected cetaceans, and the milk glands which he found convinced him that these were mammals, despite their living in the sea. He was struck by the underlying similarity of forms in different species, and his *Histoire de la Nature des Oyseaux* (1555) showed the human skeleton beside that of a bird, and became a seminal image of comparative anatomy. Leonardo had drawn sketches indicating some fundamental similarities in the skeletons of man and horse, but of course his ideas remained unknown. Belon was the first to subdivide the world of birds systematically, suggesting such major groupings as raptors, nocturnal and diurnal, web-footed birds, waders, field-birds etc. A great traveller in Europe and the Middle East, and an original observer, Belon still regarded it as part of his function to preserve the comments of Aristotle and other authorities on the animals he describes. In the same year as Belon's study of birds there appeared the most monumental of the new zoology encyclopedias, the *Historia Animalium,* 1555, by Konrad Gesner. Gesner was a prolific intellect: physician, linguist, classical scholar, botanist and one of the first palaeontologists. His work on animals was, at 4,500 pages, so massive that it was inevitably a compilation rather than an original treatise, and he still included numerous fanstasy-creatures,

Skeleton of man and bird from Belon's *Histoire de la Nature des Oiseaux,* 1555. The point of Belon's drawing is to illustrate the general similarity of form between two apparently remote species, but he could not explain the relationship.
The British Library 444.1.2.

Der viertzehend teil von

Von der grossen Meerspinnen.

Maia.   Ein Meerspinn/ Spiegelkrab/ Hechelkrab/
Mäterkrab.

Von seiner gestalt.

Je so von der natur der Wasserthieren geschriben habend/ sind in der gegenwirtiger Kraben mit einhellig. Dann D. Rondelet vermeint Maiam seyn den Krab/ welches figur hie bey anfang gesetzt ist/ auff schlichten sonderlichen ursachen/ allein von der grösse genommen. Doctor Cunrat Gesner sampt vilen anderen gelerten männeren/ vermeint Maiam cancrū zu seyn den Krab so hie nach gesetzt/ im von Venedig für das weyblin deß geschlächts/ geschickt/ gantz sauber vñ wol abconterfetet. Sölchen setzt hernach D. Rondelet für das mer Meerspinn/ das ist/ pro Paguro/ das männlin von sölchem geschlächt. Sölch Krabi sind gantz änlich einer Spinnen/ als daß die gestalt wol bezeuget: auß den sach sy Meerspinnen genennt werden/ ob gleych einer auß den Kunstfischen sölchen nammen bekompt.

Grosse Meerspinn weyblin.

Die figur ist zu Venedig abconterfetet.

Jse figur als oben gehört/ ist von Venedig kommen dem hochgelerten herr Doctor Cunrat Gesner zugeschickt durch einen seiner besten freunden/ Ich auch sölcher in Narbonensi Gallia gar offt gesehen/ ungefarlich einer span breit/ sunst garnach rond/ gentzlich in sölcher gestalt als hie erzeigt wirt.

Allerley Wallfischen.

Ser Meermünch sol sich an dreyen orten erzeigt/ an dreyen orten gefangen seyn worden. Erstlich in Nortwegia/ bey Dietz/ bey der statt dem Elpoch. Demnach sol er auch in dem Baltbischen Meer gefangen seyn worden/ bey der statt Elboa/ so 4. meyl ligt von Coppenhaga/ der houptstatt des Danischen reychs. Die gantze lenge des fisches 4. ellenbogen/ sol dē künig zugeschickt gewert/ vñ zu einem wunder behalten seyn worden. Sol von den fischeren im garn mit den Häringen gefangen seyn worden.

Dergleychen sol auch einer bey Burdegal in dem Gallischen Meer gefangen seyn worden.

Albertus schreybt/ daß dise art der fischen im Britanischen Meer seye gefangen worden.

Von dem Meerbischoff.

Episcopus marinus.   Ein Meerbischoff.

Von seiner gestalt/ vnd an welchen orten er gefangen.

Uff das jar als man zalt 1531. sol ein sölcher fisch mit sölcher gestalt gentzlich aller zierden eines Bischoffs änlich/ an dem gstad des meers der Polands nächst gefangen seyn worden/ vnnd dem Polonischen künig fürgetragē. Welches durch etwas zeichen/menschlich bedunckt wöllen bedeüten vnnd begeren/daß es ein grosse begird habe wider in das meer. Zū welchem also es gefürt ist worden/ sol es sich zū stund darzugeworffen/vnd in die tieffe verschloffen haben.

Von einem anderen Meerwunder
auff einer tafel oder zädel im Teütschland getruckt.

o iij

A crab and a monkfish from Gesner's *Historia Animalium*, 1558. Gesner's monumental work faithfully illustrated thousands of species, but also included many mythical and fabulous creatures.

The British Library 460.e.4.(3)

but it expresses admirably the new desire for completeness and for classification. All these biological pioneers were feeling their way towards Leonardo's credo, that nature functioned through rules, had its own logic, and that beneath the multiplicity of life-forms, lay common patterns and structures; yet the key to unveiling them lay still beyond their reach. The experimental method, the readiness to observe over long periods, and to draw conclusions uninfluenced by conventional doctrine, had not yet dawned. Consequently certain fundamental principles still eluded these naturalists, such as the sexuality of plants, or the very division between the plant and animal kingdoms: Jerome Bock for example, puzzled by the propagation of orchids, suggested that they were the offspring of birds, while John Gerard's famous herbal, the *Historie of Plants,* 1597, repeated the same idea, that Brent Geese were hatched from barnacles. The same credulity was evident in human biology, for travellers' tales from the New World showed that medieval belief in giants, monsters and half-human species was still very much alive at the height of the Renaissance. The printing of encyclopedic works such as those of Fuchs or Gesner served to delight the eye and the mind, and to raise the awareness of botany and zoology; but the reasoned analysis of biological form and function lay still in the future.

The empirical revolution in the life sciences is seen most clearly in human anatomy, where the skills of the artist and the printer served the cause of scientific innovation. Andreas Vesalius's great work, *De humani corporis fabrica* of 1543, inaugurated a new era of anatomical research, its text advocating first-hand, critical examination of the body's mysteries, and its illustrations offering splendid models for study. Flemish by birth, Vesalius studied at Louvain and Paris, and at the age of twenty-three was teaching anatomy and surgery at the university of Padua, where, most unusually, he performed dissections in person, a task that had

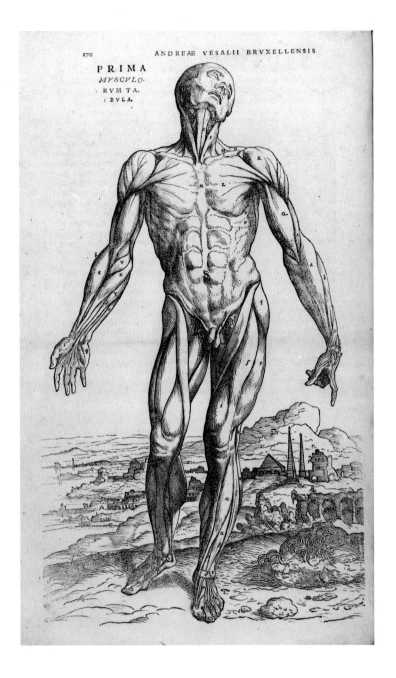

The frontal muscles from
Vesalius's *Fabrica*.
The British Library C.54.k.12.

traditionally been performed by an assistant while the master lectured. Already
Vesalius had rejected the traditional divisions between medicine and anatomy.
Orthodox, Galenic medicine regarded disease in a philosophical light, as a dis-
function of certain imagined systems, to be corrected by diet or drugs; hence the
anatomical study of organs was irrelevant, and the practice of surgery was beneath
the dignity of the philosopher-physician. Vesalius's researches, which were aided
by the sympathetic authorities in Padua who supplied him with the bodies of
executed criminals, led him to the conviction that 'Galen himself never dissected
a human body', and that his ideas of human anatomy had been based on systematic analogies with that of dissected animals, analogies which were frequently

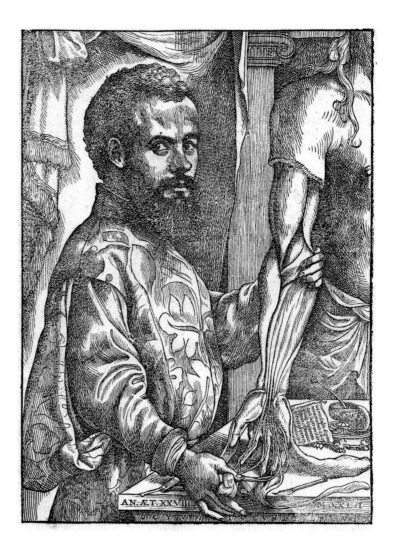

Portrait of Vesalius from
De humani corporis fabrica,
1543.
The British Library C.54.k.12.

misleading. Galen had taught that the body's central vein, the *vena cava,* had its source in the liver, while Vesalius saw it coming from the heart. Galen asserted that the human jaw is formed of two bones, which Vesalius showed is true in many animals but not in man. In discussing the brain, Galen had taught that sensation and motion were guided by the 'animal spirit' gathered in a network of arteries at the base of the brain, the *rete mirabile,* yet Vesalius plainly saw that this network did not occur in man. He also saw that nerves were not hollow vessels along which some vital fluid flowed, as Galen and all other physicians had taught.

The diagrams in pre-Vesalian anatomy books had virtually no basis in observation, but were drawn to illustrate pre-conceived schemes of physiology, showing structures and connections between organs as described in the accompanying text. Vesalius felt impelled to embody his new, observational approach and the discoveries to which it had led him, in a new reference book which would be of unprecedented scope and accuracy. He composed a thorough text describing the skeleton, muscles, nerves and internal organs, sought out the finest artists and wood engravers to create a series of superb illustrations, and supervised in person the printing of the book in Basel. Vesalius's work was by no means free of error, in particular his inability to progress beyond Galen's physiological teachings led

87.

Serratura.

The reality of medical
treatment in the age of
Vesalius. From
Gersdorff's *Feldbuch
der Wundartzney,* 1540.
The British Library C.31.m.12.

him to see the heart as comprising two chambers, which he believed must be linked by minute valves, but he stressed that he was unable to find them. His greatest weakness was in embryology and the female reproductive system, which he explained was entirely due to his having ever dissected only three females. He declined to speculate whether the heart or the brain was the seat of the soul, disdaining to become involved in controversy with the church.

Although Vesalius cannot be credited with any single overwhelming discovery, his career was revolutionary. For the first time an ancient scientific authority – the Galenic tradition which had been dominant for 1500 years – had been tested against a body of empirical facts, derived from first-hand study, and that authority had been found badly wanting. The personal, observational basis of Vesalius's text is evident in his description of many features, such as the small bones in the ear, of which he says 'the one somewhat resembles the shape of an anvil, and the other resembles a hammer', terms which are still used for those bones in the Latin forms incus and malleus. Vesalius's monumental illustrations stand at the beginning of a new form of recorded, observational science, in which later study and deduction are made possible by the availability of the scientific image. Although the defenders of authority, the Galenists, resisted for many years, Vesalius inspired many distinguished followers in Italy such as Eustachi, Falloppio and Fabrici, who together comprised a new school of empirical anatomy: where their observations contradicted Galen, they learned to rely on observation alone. Would Vesalius have been led to revise Galenic physiology and medicine had his academic career been longer? That scourge of Renaissance Europe, the plague, demanded a new model of disease as the invasion of the body by an objective outside agent, as Paracelsus had argued, and the authority of the Galenic theory of humours could no longer account for the transmission of new infectious diseases such as syphilis. However, in the year of the publication of the *Fabrica,* Vesalius turned suddenly from university teaching to become court physician to the Emperor Charles V, attending gluttonous princes, syphilitic courtiers and battlefield casualties for ten years. In 1564 when he was about to resume his former chair at Padua, Vesalius died in rather miserable circumstances on the island of Zakinthos, where he had been shipwrecked while en route to Palestine. The site of his grave is unknown.

GEOGRAPHICAL SCIENCE

In the context of Renaissance history, the applied science which emerged most naturally was navigation. Mediterranean seafaring had a history of several thousand years, during which her pilots relied mainly on dead-reckoning and coastal sailing, techniques which in the pre-Christian era enabled traders to sail from Phoenicia to Britain. By the late thirteenth century, the magnetic compass had arrived in Europe (whether independently invented or imported from Arabia or China has never been established), and accurate charts with compass lines were being drawn, which helped enormously when sailing out of sight of land. But position-finding as a science did not exist: a course was set, maintained with the help of the compass, until familiar coastal features were sighted. This was acceptable in the Mediterranean, with traditional skills built up over years. But when first the Portuguese then the Spanish ventured into the oceans beyond the Pillars of Hercules, position-finding became a matter of life and death. When Vasco da Gama was en route for India in 1497 he spent some 96 days at sea, until then the longest European voyage, out of sight of land. Clearly spherical astronomy must be invoked to determine position in circumstances like these. The theory was not complex, in respect of latitude at least, and had been taught even in medieval works of astronomy such as Sacrobosco's *De Sphaera,* where the elevation of the Pole Star above the horizon is the observer's latitude. But of course the Pole Star vanishes at the equator, and in the absence of a southern equivalent, latitude can be gauged from the Sun, but this is complicated by the Sun's shift in elevation through the year, and tables must be used. A mariner's astrolabe, a circular scale with a sighting device, was used to 'shoot the Sun' at the meridian. By this means the Portuguese, en route for the Cape, learned to sail with confidence far out into the Atlantic in order to pick up favourable winds, then on reaching the desired latitude, they turned directly east and 'sailed down the latitude'. Columbus too had reasoned that if he sailed due west down the latitude of 28 degrees north from the Canaries, he must reach the coast of Cathay.

Longitude was a much tougher problem, because unlike latitude it has no objective marker in the sky, for the celestial sphere is constantly revolving in the

A navigator sighting the altitude of the Pole Star to determine his latitude, from Medina's *Regimiento de Navegacion*, 1563.

The British Library C.125.b.4.

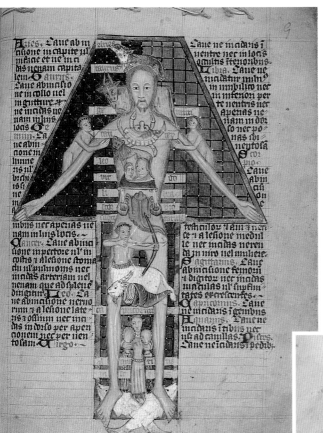

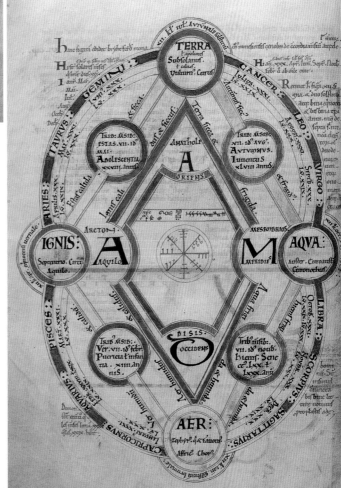

One of the central principles of medieval medicine was that the parts of the body are ruled by the signs of the Zodiac, from Aries ruling the head, to Pisces the feet. Blemishes or diseases of those parts were considered to be the result of astrological conjunctions.

Bodleian Library, Oxford

Cosmic scheme: this diagram illustrates perfectly the medieval desire to see the phenomena of the world as forming a harmonious whole. The four elements, the four cardinal directions, the seasons, the four ages of man, the twelve months and signs of the Zodiac, the twelve winds, the equinoctial and solstitial points – all are woven into a symmetrical pattern.

St John's College, Oxford

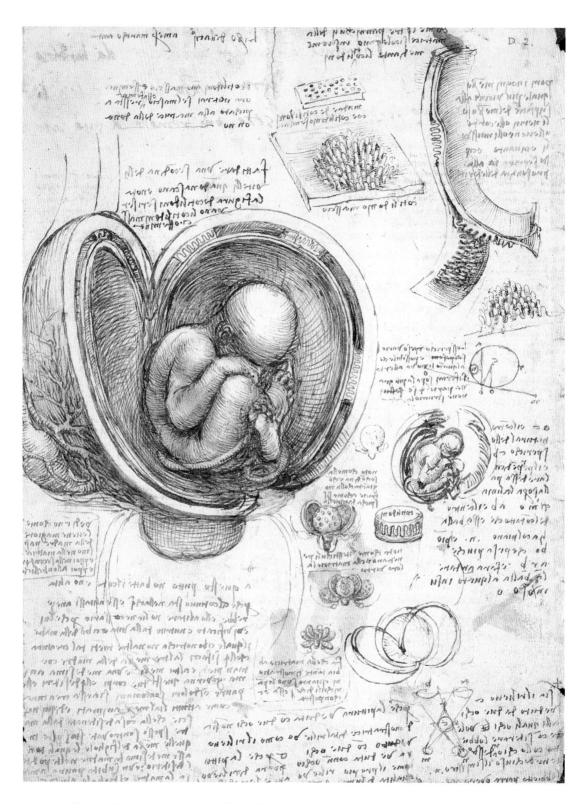

Human foetus by Leonardo da Vinci. The notes that surround this powerful drawing tell us
that the foetus was four months old. Leonardo records his mystification over the origin the soul,
some time between conception and birth.

Royal Library, Windsor Castle

plane of longitude. It was correctly understood to be a function of time, each fifteen degrees of longitude representing an hour's difference in the positions of the heavenly bodies; but the absence of accurate clocks which could function at sea meant that this principle was of little practical use. This level of navigational science was not common among seamen, and it was some years before theoretical works on navigation began to appear. Several treatises by the foremost Portuguese mathematician, Pedro Nuñez, must have been beyond the reach of all but an elite few. More accessible was the work of the Spanish cosmographer Pedro de Medina (1493–1576) whose *Arte de Navigar* was published in 1545, and was translated into French, English, Italian and German, and became the basis of many other navigational works, such as *The Seaman's Secrets* by John Davis (of Davis Strait). Interestingly, Medina opens with a description of the structure of the universe which is purely Ptolemaic, and this is copied by Davis in his version, published in 1595. Medina's treatise contains one significant omission – its failure to remedy the inaccuracy of the contemporary sea-charts, which were drawn without any mathematical structure, or projection. Sixteenth-century charts were drawn on a simple squared basis, as if the ratio of latitude to longitude were constant, which of course it is not. In effect these charts ignored the sphericity of the earth: a compass line on such a chart of the Atlantic or Indian Ocean would cut the meridians always at the same angle, but this line when plotted on a globe becomes a curve, not a great circle. This defect would later be remedied by the great Flemish mapmaker Gerard Mercator, but its persistence in the late sixteenth century demonstrates that navigation and cosmography were reacting rather slowly to the demands of the age, and that although a scientific basis for navigation existed, communication between scholars and seamen was lacking. The belief that navigators in the age of exploration contributed to a revival of astronomy is a the reverse of the truth.

COPERNICUS

It was within astronomy that there occurred one of the most profound events in the whole history of scientific thought: the Copernican revolution. The replacement of the earth-centred, finite universe, accepted probably by all men throughout recorded history, is one of a handful of major landmarks in man's intellectual development that have fundamentally shaped our entire understanding of our world and ourselves. Yet the role of Nicholas Copernicus (1473–1543) in the revolution associated with his name is, on close examination, more ambiguous than is often supposed. The natural model of a scientific revolution is to imagine that scientists who had worked within an accepted intellectual framework were confronted with new data which proved irreconcilable with that framework, so that they were driven to formulate radical new theories. Yet nothing like this happened in the case of Copernicus: new facts, new astronomical observations and new evidence were all strikingly absent from his work. He was a student of books rather than nature, and observation was not the basis of the new theory. Instead his achievement was to perform a highly original 'thought-experiment' in devising a new geometric model which accounted for the movements in the heavens in a simpler, more elegant way than the Ptolemaic model did. Nor does Copernicus exactly fit the image of a revolutionary scientist: a churchman with a very traditional education in canon law and a private interest in astronomy, who spent his life in ecclesiastical administration, and chose not to publish his great theory for some thirty years after its conception. Copernicus left no detailed

Portrait of Copernicus; anonymous and of unknown date, it was once considered to be a self-portrait.
Museum Okregowe, Torun

account of the genesis of his new vision of the heavens, but we know from his own published statements that he was dissatisfied with Ptolemy's geometry, whose complex system of eccentrics and equant points seemed impossible to

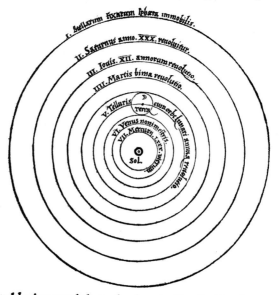

NICOLAI COPERNICI

net, in quo terram cum orbe lunari tanquam epicyclo contineri diximus. Quinto loco Venus nono mense reducitur. Sextum denicʒ locum Mercurius tenet, octuaginta dierum spacio circū currens. In medio uero omnium residet Sol. Quis enim in hoc

I. Stellarum fixarum sphæra immobilis.
II. Saturnus anno. XXX. reuoluitur.
III. Iouis. XII. annorum reuolutio.
IIII. Martis bima reuolutio.
V. Telluris cum orbe lunari annua reuolutio.
VI. Venus nonimensis.
VII. Mercurij LXXX dierum.
Sol.

pulcherimo templo lampadem hanc in alio uel meliori loco po neret, quàm unde totum simul possit illuminare? Siquidem non inepte quidam lucernam mundi, alij mentem, alij rectorem uo cant. Trimegistus uisibilem Deum, Sophoclis Electra intuentē omnia. Ita profecto tanquam in solio regali Sol residens circum agentem gubernat Astrorum familiam. Tellus quoqʒ minime fraudatur lunari ministerio, sed ut Aristoteles de animalibus ait, maximā Luna cū terra cognationē habet. Concipit interea à Sole terra, & impregnatur annuo partu. Inuenimus igitur sub hac

The Sun-centred cosmos, from Copernicus's *De revolutionibus,* 1543.
The British Library 59.i.6.

reconcile with the movements of real, physical spheres, and with the concept of uniform circular motion. These were objections which had a long history in Islamic and western astronomy. While seeking various ways out of the impasse, Copernicus took the crucial step of considering that the apparent movements of the heavenly bodies might be an effect of the movement of the observer on earth:

> ' A seeming change of place may arise from the motion of either the object or the observer. ... If then some motion of the Earth be possible, the same will be reflected in external bodies, which must seem to move in the oppo-site sense ... I began to think of the mobility of the Earth; and though the

opinion seemed absurd … I considered that I might be allowed to try whether by assuming some motion of the Earth, sounder explanations for the revolution of the celestial spheres might also be discovered.'

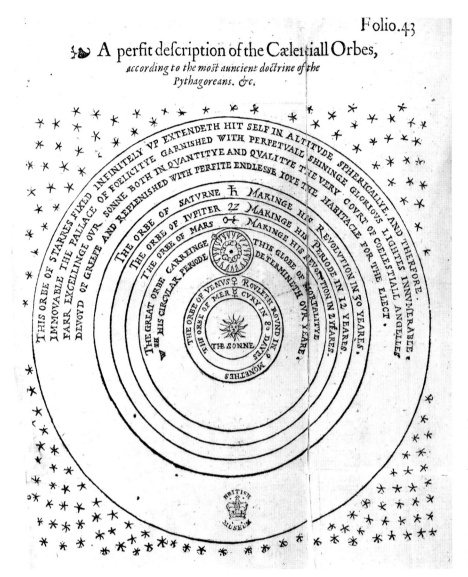

The first English account of the Copernican system, in Thomas Digges's *Prognostication Everlasting*, 1576. This version is important for its explicit statement that the sphere of the stars extended 'infinitely up', and that the stars were suns.
The British Library 718.g.52.

And so it proved, for working first from the movements of Mercury and Venus, Copernicus realised that their constant proximity to the Sun was much more easily explained by supposing that they revolved around it and not around the Earth. But Copernicus accepted the physical reality of the celestial spheres, hence it was impossible that these two planets should orbit the Sun, while the Sun and other planets still orbited the Earth, for the spheres would intersect. Only by placing the Sun at the centre of the system, and the Earth in motion around it could all the celestial spheres be preserved, and in the same order proposed by Ptolemy, only reversing the positions of the Sun and the Earth. In working out the detail of this theory, Copernicus still had to account for the irregularities in the planets' orbits which spring from the fact they are not uniformly circular. He did this in the traditional Ptolemaic way, by inventing more epicycles, an example of his purely mathematical approach to this problem. As Galileo was later to point out, the epicyclic path ascribed to Venus by Copernicus could not account for the planet's constant brilliance, showing that empirical observation was not part of Copernicus's method. His revolution was a purely conceptual one, and not a 'discovery', for he was never able to offer any *evidence* of its validity.

Copernicus's theory was essentially complete by around 1510, and within a year or two of that date he had circulated a brief account of it in manuscript among a few friends. It is difficult to believe that in the long years that followed, before his death in 1543, Copernicus was not pondering the far-reaching implications of his ideas. There is one very revealing detail concerning his sources and his attitude to what he had done: in his working manuscripts, Copernicus mentions Aristarchus, the Greek mathematician famous for having proposed a heliocentric universe. Yet this reference was cancelled before Copernicus's book was published, because Copernicus recalled that Aristarchus's contemporaries had wished to indict him for impiety because of his outrageous views. It is not certain if Copernicus would ever have published his theory, had not his hand been forced by the printing in 1540 of an account of his system by an Austrian mathematician, George Rheticus, whose meetings with Copernicus may thus be said to have precipitated the scientific revolution. Rheticus's *Narratio Prima* ('First Report') was the first public announcement of the Copernican system, and the absence of any explosive reaction may have led Copernicus to prepare for the press the long-withheld manuscript of *De revolutionibus orbium coelestium* ('On the Revolutions of the Heavenly Bodies'). The book was printed in Nuremberg in 1543, but the famous story that a copy was placed in Copernicus's hands as he lay on his deathbed on the May 25th of that year is, sadly, almost certainly untrue. By a quirk of fate, the printing was overseen by an astronomer with whom Copernicus had corresponded, Andreas Osiander. Osiander was also a theologian, and he was shocked by the novelty of the book, and, without any authorisation, inserted an anonymous preface claiming that the theory it contained was a purely mathematical model, and was not intended to represent physical reality. That this preface was not the work of the author was unknown until the following century, and may have served to deflect immediate criticism, for many years were to pass before Copernicus's theory was widely accepted, or its profound implications fully understood.

What were these implications? The first was a set of problems in physics relating to the earth's movement, some being commonsense, others more philosophical. If the earth was whirling through space, why was everything on it not flung violently away? Why did a stone thrown directly into the air not fall some distance away from its launching-point? Questions like these could be answered by pointing out that everything on the earth shared the earth's motion, just as an object dropped from the mast of a ship will imediately below the mast, and not be left behind as the ship sails on. More troubling was the challenge to the central doctrine of physics established since Aristotle, that falling objects are seeking the centre of the universe, located at the centre of the earth, and that all natural motion is to be explained in this way, except the spherical motion of the heavens, where other elements and other laws exist. But if the earth is not the centre of the universe, why do heavy objects fall downwards at all? In the Aristotelian universe, all the heavy elements – earth and water – naturally gather at its centre, while air and fire rise upwards; but if the earth is not the centre, this cannot be true. Copernicus suggested instead a principle of cohesion, in which all

Pantheon of astronomers from Cellarius's *Atlas Coelestis*, 1660. Copernicus, holding a model of his system, is clearly recognisable; Tycho is on the left and Galileo on the right. The bearded, turbanned figure is Ptolemy, the other possibly Ulugh Beg. The figure in black may be Cellarius himself, while the notable omission is Kepler.

The British Library Maps C.6.c.2

heavy matter would indeed gather into a sphere, which need not however be the centre of the universe. But if the earth is a planet, may not the other planets also be earths, and the age-old division of the cosmos into two realms be a fable? If the massive earth is revolving in space, what possible power keeps it there, and maintains the entire system in equilibrium? The second profound implication of Copernicus's theory concerned the scale of the universe, and was equally destructive of traditional thought. If the earth is circling the sun, the diameter of its orbit must be vast, to be measured in millions of miles; in that case the well-known phenomenon of parallax should produce large and regular variations in the positions of the stars through the course of the year. But no such variations had ever been observed, hence indeed the term 'sphere of the fixed stars'. The only possible inference – if Copernicus were correct – was that the stars were so distant that even a movement as great as the annual orbit of the earth could produce no parallax. If this were true, then the scale of the universe must be immeasurably greater than anyone had previously conceived, nor was there any longer any reason to suppose that the Sun lay at its centre. The world of man had been displaced from being the focus of the whole creation, to being a random point in an immense and now uncharted universe. Copernicus was aware of this, and he expressed it by saying that lines drawn from a star to any point on the earth's surface would be parallel to each other. 'The heavens', he concluded, 'are immense, and present the aspect of an infinite magnitude'. There was a long medieval tradition of speculation that the universe was infinite, and with Copernicus this idea became a distinct physical possibility.

De Revolutionibus was a work of fairly advanced mathematics; its author had apparently conceived it only as a theoretical exercise, and in any case he was dead; many intellectuals rejected the new theory as absurd, and no school of followers sprang up immediately to defend it. These factors may explain why reaction to the theory was somewhat delayed, and why the Church, deeply engaged in its campaign of Counter-Reformation, failed to grasp its full implications. Copernicus stands at the end of a scientific tradition of purely mathematical astronomy, for his work was conceptual and not empirical. In view of the pervasive climate of innovation in Renaissance thought, it is perhaps surprising that this great intellectual revolution should occur not in a new, empirically-based science such as biology, chemistry or geology, but in a sphere of the exact sciences with a tradition of thousands of years behind it. Innovation in the Renaissance was predominantly concerned with the things of the real world – with architecture, navigation, printing, cartography, mining and so on – and not with fundamental concepts. After Copernicus's death, among those who brought out most fully the revolutionary vision latent in his work, were the mystics and magicians such as Giordano Bruno, whose belief in the infinity of the universe and the multiplicity of Suns and planets deeply alarmed the Church, for it implied that the religious view of this world as the unique theatre of God's activity, was false.

In the history of science, the problem of the Renaissance is the problem of a mismatch between society and learning: society was dynamic, learning was static, even backward-looking. In the universities the teaching of logic, mathematics or theology proceeded as it had done for centuries, and the printing press did not in itself incite men to print new ideas. The universe was seen through Ptolemy's eyes, the human body through Galen's, theology through those of Aristotelian logic. This is easily seen if we consider the world-view of Shakespeare or Marlowe, who could assume in their audiences a belief in hell and purgatory, in the Ptolemaic universe, in the four elements, in witchcraft and divination, in the celestial spheres and their music. Marlowe's Faustus is the archetype of the

Renaissance magus, obsessed with controlling nature before he has understood it.
The importance of mystics and magicians like Bruno and Paracelsus is their clear
desire to break out of the sterile world of academic learning, into a more personal
and powerful understanding of nature. They produced systems of philosophy
radically different from the natural philosophy which had developed over centuries within Christian theology, and for this they have been claimed as prophets
of the scientific revolution. But any direct line from their speculations, from the
occult correspondences which they found in nature, to empirical science as it later
developed, is very hard to trace. Nature had been rediscovered certainly, but not
measured or analysed. The painter, the navigator, the mining engineer, the naturalist, the mapmaker, the magus – all were exploring the diversity of the natural
world, but the language which would bring order into that diversity still awaited
discovery.

Chapter Five

SCIENCE REBORN

&❧

'This is the foundation of all: we are not to imagine
or suppose, but to *discover,* what nature does or
may be made to do. '

Francis Bacon

THE ASTRONOMERS

If any one date marks the beginning of the scientific revolution, it is surely 11
November 1572. Shortly after sunset on that day, the Danish astronomer Tycho
Brahe glanced up at the constellation Cassiopeia, and was astonished to see a star
far brighter than any of those surrounding it, a star which he instantly realised
had not been there before. Hurrying home, he measured its position with sighting instruments, a process which he repeated on many succeeding nights, despite
the gradual lessening in its intense brightness, until it vanished in March 1574.
This remarkable celestial body had no tail, and was therefore no comet; it did not
move as a planet does, nor was the slightest change in its position observed over
the seventeen months of its visibility. Tycho concluded that it was indeed a star, a
new star which had flared into intense life, and then faded back into darkness. The
implications of this event for traditional cosmology were profound, for every
scholar in Europe accepted as sacrosanct the Aristotelian doctrine that the heavenly spheres and all they contained were fixed and immutable, while generation,
change, and decay occurred only on earth. Far from being merely a concept in
pagan science, this belief had of course been strongly reinforced throughout the
Christian middle ages by religious ideas of heaven as the dwelling-place of God
and the angels.

Tycho described the new star (which we now know to have been a supernova)
in a short treatise *De nova stella,* 1573. The question-mark thus raised over traditional cosmology became even starker a few years later, with the appearance in
November 1577 of a great comet, whose position Tycho observed nightly for
two months. He concluded that its path lay not below the moon, the prevailing
belief at the time, but between the moon and Venus. It is often said that it was
this comet which caused Tycho to abandon belief in the reality of the celestial
spheres; this was indeed the position that Tycho arrived at, but he did so some
years later, after a prolonged study of all the planetary orbits, and of a second
comet which appeared in 1585. It was as a result of these observations that Tycho
proposed his own theory of cosmic structure, which differed from that of
Ptolemy and of Copernicus. Tycho became convinced that the planets did orbit
the Sun, as Copernicus had suggested, but he could not accept the reality of a
moving earth, principally because none of his own painstaking observations had
ever revealed the slightest parallax in the stars. Therefore he proposed that the

Tycho's new star: the
observation of a new star
(now known to be a
supernova) in the constellation Cassiopeia in 1572
was a turning-point in the
history of astronomy,
overthrowing the age-old
belief that the heavens
were perfect and
unchangeable.
The British Library C.54.d.15

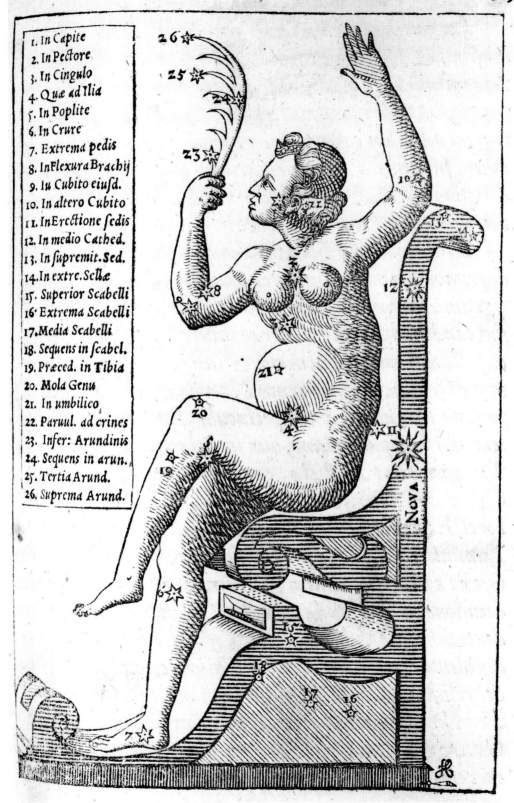

1. In Capite
2. In Pectore
3. In Cingulo
4. Quæ ad Ilia
5. In Poplite
6. In Crure
7. Extrema pedis
8. In Flexura Brachij
9. Iu Cubito eiusd.
10. In altero Cubito
11. In Erectione sedis
12. In medio Cathed.
13. In supremit. Sed.
14. In extre. Sellæ
15. Superior Scabelli
16. Extrema Scabelli
17. Media Scabelli
18. Sequens in scabel.
19. Præced. in Tibia
20. Mola Genu
21. In umbilico
22. Paruul. ad crines
23. Infer: Arundinis
24. Sequens in arun.
25. Tertia Arund.
26. Suprema Arund.

NOVA

earth was still the unmoving centre of the universe, around which the Sun revolved, but that the other planets were orbiting around the Sun. In this scheme, the orbit of the Sun crossed those of Mercury, Venus and Mars, hence necessitating the abandonment of any belief in the reality of the heavenly spheres. The Tychonic system fitted the observed paths of the planets as well as the Copernican system: the choice as to which one was true was largely determined by belief in the possibility or otherwise of a moving earth.

This break with traditional cosmology was dramatic enough, but in historical perspective what is equally important was its empirical basis: Tycho broke away from classical theory, from purely geometric analysis, into a new model based on observation and measurement. In this he contrasts with Copernicus, the origin of whose theory had been purely conceptual, and unlike Copernicus, Tycho was consciously setting out to reform the practice of astronomy. His 'compromise position', between Ptolemy and Copernicus, proved popular, restoring the earth to the centre of the universe, avoiding both the objections to a moving earth, and the possibility of conflict with the church.

Securing the patronage of the Danish crown, Tycho (1546-1601) established on the island of Hven (now Ven, in the Danish Sound) an observatory equipped with the largest and most accurate instruments yet built in Europe. Between 1576 and 1597, he and his many assistants (among whom were Willem Blaeu, the great cartographer, and later Johannes Kepler) laid the foundations of a new era in astronomy. He re-calculated the positions of the thousand-plus Ptolemaic stars, positions he marked on a huge celestial globe five feet in diameter, and which he published in a new star catalogue. His thousands of careful observations of planetary positions led not only to his new theory of cosmic structure, but, in the hands of Johannes Kepler who inherited this data, to a new approach to celestial mechanics. For the abandoning of belief in the celestial spheres posed in even sharper form the question raised by Copernicus's moving earth: what force is it that sustains the cosmos, that moves the earth and the other planets? Whatever the religious beliefs of a man like Tycho, there is not the slightest doubt that within a few years of his death in 1601 the answer to this question would be sought in what we would now recognise as scientific terms. Few men doubted that God ruled the universe, but a growing elite of scholars were to be found in almost every country in Europe who believed that he did so in ways capable of being analysed by mathematics and human reason. Many in this elite were convinced Copernicans or Tychonians, and were conscious of having severed themselves from the dead-end of Aristotelian physics, just as the followers of Vesalius were consciously anti-Galenic. It was the French pedagogue Pierre de La Ramée who boldly claimed in the 1540s that 'everything Aristotle said is false'.

If Tycho played a vital role in inaugurating an empirical revolution in astronomy, Kepler was arguably a still greater figure, for he constructed the first physical theories to fit the new cosmos described by Copernicus and Tycho. Kepler (1571-1630) was a deeply religious thinker, constantly seeking the underlying harmonic structure which he believed God had built into the universe, but he was also a tenacious and brilliant mathematician, who spent the greater part of his life in the almost interminable calculations which he believed would reveal those structures. His earliest and most celebrated cosmic theory was conceived when he was but twenty-five years old, and related the orbits of the planets to the five regular solids of classical geometry. Meditating why there should be just five planets, and why their orbits were spaced as they were, Kepler was inspired with a geometric vision of those orbits inscribed within the five regular solids of classical geometry – the cube, the tetrahedron, the octahedron, the icosahedron, and the dodecahedron. This intriguing model, which seemed to revive the Pythagorean doctrine that nature is fundamentally mathematical, does indeed approximate to the truth; but it transpired to be an accidental relationship, unrelated to any laws of science. The treatise in which Kepler set out this theory, the *Mysterium cosmographicum* of 1596, was the first avowedly Copernican work since *De revolutionibus*, for it was entirely based on the concept of the Sun-centred universe. 'I wanted to become a theologian', Kepler wrote, 'and for a long time I was restless. Now however, behold how God is being celebrated in astronomy.'

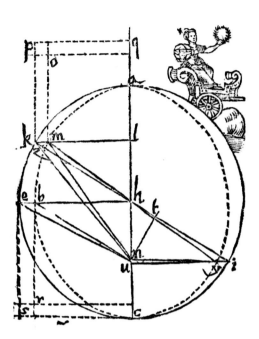

Unlike Copernicus or Tycho, Kepler could not rest with geometric descriptions of the cosmos, but was impelled to seek its driving-force. He felt instinctively that the Sun was more than simply the geometric centre, and was in some way the focus of a force system. He believed that the orbital periods of the planets reflected the diminution of this force, hence the more distant a planet was from the Sun, the longer its period. Thus Kepler was the first astronomer to consider celestial mechanics as an alternative to the now-discredited heavenly spheres. Aware of William Gilbert's assertion (see below p144) that the earth acted as a great magnet, Kepler explored the idea that the cosmic force might be magnetic, emanating from the Sun, and diminishing with distance; thus in a sense he arrived at the threshold of a gravitational theory.

During the next twenty years, Kepler worked incessantly with Tycho's data to build up a mathematical description of the solar system. It became apparent to him that the doctrine of uniform circular motion was false, that the planetary orbits were not circles, and that the irregularities of their paths and speeds were not to be explained by inventing epicycles. Although at first unable to characterise these paths accurately, he recognised that the planets accelerate as they approach the Sun, and slow down as they move away from it. He expressed this brilliantly in kinetic terms by saying that the radius vectors from Sun to planet sweep out equal areas in equal times, a formula which became known as Kepler's second law of planetary motion. The clear implication of this law was that the Sun *controls* the motion of the planet. The analysis of the paths of the planets cost Kepler enormous trial and error before he arrived at his first law, that they are ellipses, with the Sun at one focus; with this discovery, Kepler swept away the complex epicyclic system which had prevailed since Ptolemy. Some years more were to elapse before he formulated the third and most powerful of his planetary laws, that the orbital periods of the planets have a precise relationship to their distance from the Sun. This harmonic law is expressed by the formula $P^2 = a^3$ where P is the planet's orbital period in years, and a is its distance from the Sun in Astronomical Units (i.e. the distance from earth to Sun). Thus Jupiter is 5.2 AU from the Sun; 5.2^3 is 140.6, and the square root of 140.6 is 11.8, which is Jupiter's orbital period in years. The power of this law is such that the distance of any body in motion about the Sun can be calculated by observing its orbital period.

Kepler had brought astronomy into a new age, and his work was an essential prelude to that of Newton. His quest for the harmonic laws of the universe had in it an element of mysticism, but, unlike the mysticism of the hermetic school,

Kepler's analysis of the orbit of Mars. Kepler spent seven years analysing the daily positions of Mars recorded by Tycho, and concluded that its orbit was not a perfect circle as traditionally believed, but an ellipse, shown here by the broken line. The Sun is at *n*, exaggeratedly offset from the ellipse's centre. The figure bearing the laurel crown symbolizes Kepler's triumph in reforming planetary astronomy.
The British Library C.133.h.3.

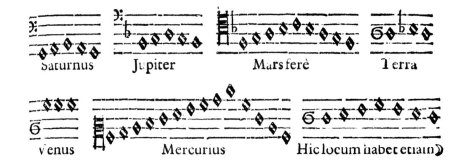

Saturnus   Jupiter   Mars ferè   Terra

Venus   Mercurius   Hic locum habet etiam ☽

Cosmic harmonies
according to Kepler. In a
work of 1619, *Harmonice
Mundi*, Kepler announced
his theory that the velocities of the planets are
related to musical scales,
rising and falling as they
approach and recede from
the Sun. Kepler never
ceased searching for
hidden patterns which he
believed must be built
into the structure of
the universe.
The British Library C.124.f.9.(1)

Kepler's was expressed through rigorous mathematical analysis. Kepler's life was sometimes deeply unhappy, as he was harried from city to city by the religious conflicts that plagued Germany, and his tenacity and intellectual integrity make him a towering figure. It is impossible not be moved by the epitaph he composed for himself:

Mensus eram coelas, nunc terrae metior umbras
I once measured the heavens;
now I shall measure the shadows of the earth

Printed portrait of
Galileo, 1613, looking
pugnacious and self-
confident. One of the
cherubs holds a telescope,
although the artist has
given it the appearance
of trumpet.
The British Library 8710.dd.66

The third of the founders of modern astronomy, Galileo Galilei (1564–1642), exemplifies even more clearly than Kepler the seventeenth-century transformation in the study of nature, away from philosophy and system-building and towards empiricism and mathematics. His conflict with the Church makes him appear a pivotal figure in the history of science, marking the emergence of a new force in European intellectual life. Galileo's career fell into several distinct phases, as external events channelled his energies into a succession of inquiries and conflicts. For the first twenty years of his adult life he held chairs of mathematics at Pisa and then at Padua, and his interest centred on mechanics, and on an attempt to devise a mathematical language of bodies in motion. At a very early stage he dismissed Aristotle's physics, regarding his absolute distinction between the heavenly and earthly realms as a metaphysical assumption which had been allowed to grow into a myth. Aristotle's qualities and categories and chains of logic, long considered self-evident, were to Galileo 'a bottomless ocean, where there is no getting to shore; for this is navigation without compass, star, oars and rudder'. As a convinced Copernican, Galileo was especially concerned to devise a new theory of motion, and he made lengthy studies of falling bodies and inclined planes. Whether he actually carried out the celebrated experiment of dropping weights from the tower in Pisa is unknown, but it is certain that he was striking out into new territory when he generalised his knowledge of acceleration into a law of falling bodies: that distances fallen increase as the squares of the times taken, $s \propto t$, for example 16, 64, 144 feet in 1, 2, and 3 seconds. Galileo thus identified time as an essential component in the description of motion, without which its mathematical analysis could not proceed. His Copernicanism also led him to consider very clearly the relativity of all motion: for example a man on board a ship sees an object fall perpendicularly from the mast, but a watcher on shore sees a curved line of fall, while a third observer in space sees the entire scene including the watcher also moving. Moreover Galileo had no doubt that these multiple frames of reference existed throughout the universe, that, to echo Giordano Bruno, if the earth is a planet, then the other planets are earths and the other stars are suns. Galileo thus enunciated the vital principle of the homogeneity of nature: the same fundamental laws of matter and movement must apply throughout the universe, a principle he later embodied in one of the most famous statements in the history of science:

> 'Philosophy is written in that great book which ever lies before our gaze – I mean the universe – but we cannot understand it if we do not first learn the language and grasp the symbols in which it is written. The book is written in the mathematical language, and the symbols are triangles, circles, and other geometric figures, without the help of which it is impossible to conceive a single word of it, and without which one wanders in vain through a dark labyrinth.'

It was in 1609 that Galileo turned from applied mathematics to observational astronomy, with results that were far-reaching and profound for himself and for science. He heard reports that a Dutch technician, Hans Lippershey, had combined two lenses in a device which made distant objects appear miraculously enlarged. Greatly excited – as much by its commercial potential for military and maritime use as by its scientific novelty – Galileo quickly built a telescope with a nine-power magnification, and by the end of the year he had developed one of thirty power. In January 1610 he turned his telescope to the skies: he observed the surface of the moon to be fissured and mountainous like that of the earth, and endowed with what he thought were seas; he saw that the Milky Way was composed of numerous stars; he saw that the planet Jupiter possessed four satellites, which functioned as a Copernican system in miniature; later in the year he was to observe that Saturn was surrounded by a strange oval satellite, and that Venus exhibited phases as the Moon did; perhaps most significant of all, wherever he turned his telescope, any field of stars became extended and replicated – more and more stars were revealed appearing without end in the depths of space. These results Galileo hastened to make public in a brief illustrated book, *Sidereus nuncius* ('The Starry Messenger') of 1610, perhaps the most important document of scientific reporting ever published. With this new data, man's perception of nature

Stars in Orion's belt, from Galileo's *Sidereus Nuncius,* 1610. This was the first star map produced with the aid of a telescope: Galileo found some 80 stars where the naked eye sees perhaps a dozen.
The British Library C.112.c.3.

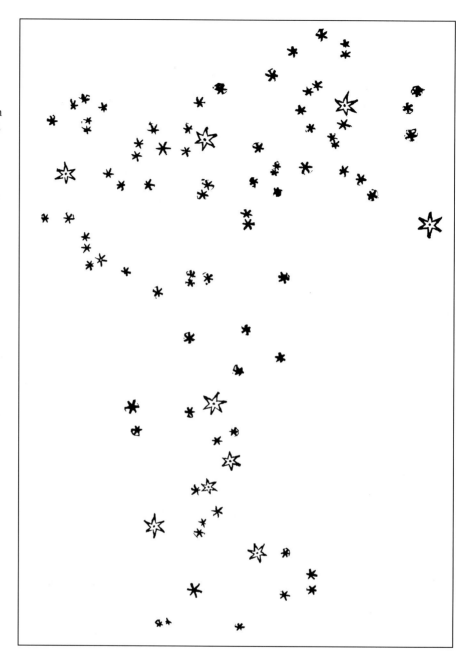

was enlarged, suddenly, at a stroke, as never before, and many long-cherished beliefs about the structure of the universe were rendered obsolete.

By the time of the publication of *Sidereus nuncius,* the Catholic Church in Italy had awakened to the theological implications of Copernicanism, and the Inquisition had come to regard it, and any unorthodox forms of natural philosophy, as punishable heresy. Galileo felt impelled to argue publicly that the investigation of nature was the province of the scientist, while the theologian's task was to reconcile science with the biblical teaching. This distinction was entirely novel in its time: it was a manifesto for the future, but it was highly dangerous to Galileo himself. Summoned to Rome in 1615, he was expressly forbidden to teach the Copernican principle that the earth moves. Thus warned, Galileo turned to what seemed less controversial topics, including the problem of longitude, and he produced in his book *Il Saggiatore* ('The Assayer', 1623) an original discussion of the scientific method, in which he rejected the acceptance of traditional authorities in matters that were subject to direct investigation. The election of a new pope, the scholar Maffeo Barberini, who had long been friendly towards Galileo, brought permission to discuss the Copernican theory in print, provided that the Ptolemaic system was given equal weight. The composition of the *Dialogue Concerning the Two Chief World Systems* occupied Galileo for six years, and came near to costing him his life. Cast in the form of a dialogue between the Copernican Salviati and the Aristotelian Simplicio, the arguments are heavily weighted in Salviati's favour, and the whole work is an attempt to demonstrate the physical reality of the Copernican system. (Simplicius was the name of an important sixth-century Aristotelian scholar, but this reference may have been lost on many readers, especially in the Italian form, and the name was readily interpreted as merely 'simpleton'). Among the arguments which Galileo employs is a carefully-constructed (but incorrect) theory of tides based entirely on the earth's dual motion, its axial rota-

Science Reborn
&❧

Christoph Scheiner, the Jesuit astronomer, observing sunspots in 1611. Scheiner was the first to announce that sunspots were surface-phenomena, and not satellites of the Sun, as was often supposed; Galileo however claimed to have made the same discovery earlier.
The British Library 532.l.6.

Immiſſione Refractoria compoſita.

Maculæ et Faculæ ex uariis obſeruandj modis, ſtabiliuntur.

tion and its orbit around the Sun. In fact much of the book focusses in this way on physics rather than on astronomy, in order to provide new theories of motion consistent with the vision of the earth as a planet, a free body moving through space. Knowing the fate of Bruno and others, was Galileo naive in advancing such views, or did he believe that, under the new pope, the moment had come to vindicate the new philosophy publicly?

If so, he was swiftly disillusioned: the response from Rome was that the book must be withdrawn, and Galileo must present himself before the Inquisition. The pope was evidently incensed that the orthodox Aristotelian position had been presented by a 'simpleton', he was apparently unaware of the earlier ban imposed on Galileo in 1616, and he felt that Galileo had defied or tricked him. The trial in

Rome in spring 1633 followed its inevitable course: Galileo was commanded, under threat of torture, to abjure the Copernican doctrine and affirm his belief in the Ptolemaic system. Who will blame him for obeying? The authorities had been too late to suppress his book, it was being read throughout Europe and he knew that he would carry the future with him, whatever lies he might be forced to utter. Sentenced to house arrest for life, he returned to his early study of mechanics, and spent his last years composing *Discourses and Mathematical Demonstrations Concerning Two New Sciences.* The sciences in question were the strength of materials, treated as a branch of statics, and kinematics, the study of velocity and acceleration. Galileo also refers to problems such as the weight of air, the propagation of sound, the speed of light, trajectory paths and the behaviour of the pendulum. Although his ideas on these subjects were not always correct, it was Galileo who introduced the principle that they were amenable to experiment and to mathematical analysis. Perhaps surprisingly, he made no contribution to celestial mechanics, although he was aware of Kepler's work, for his attention was fixed on finding a new language of physics for the Copernican age. By avowedly expelling philosophical speculation from science, Galileo became, paradoxically, the most important spokesman for a new philosophy of science.

In the post-Galileo period, astronomy continued to mature in the hands of a number of outstanding figures. Johannes Hevelius published the first detailed moon maps in 1647, and later an important new star-survey; Ole Rømer discovered that light travelled at a finite speed in 1676; Christian Huygens correctly described the rings of Saturn by 1655, and observed its satellite, Titan; in the 1670s John Flamsteed began his monumental charting of almost 4,000 star positions – the first comprehensive telescopic star survey, and the first re-mapping of the heavens since Ptolemy.

THE THEORISTS

It was the emergence of a conscious philosophy, programme or method of science among seventeenth-century thinkers that justifies our speaking of a scientific revolution in European thought. Galileo pointed the way by advocating the quantitative study of nature, in place of speculation and system-building, and the shift from verbal to mathematical analysis which he pioneered was to become the hallmark of the new science. Yet even before the revolution in knowledge associated with Tycho and Galileo, a number of sixteenth-century thinkers had begun to attack the walls of the Aristotelian fortress. The French pedagogue, Pierre de La Ramée (1515–1572), published in 1562 a proposed reform of the University of Paris, which aimed at strengthening mathematics and modernising the teaching of science: Ramée regarded Aristotle's physics as an exercise in logic, not as a description of the real world. Attracted by Copernicanism, Ramée was still critical of its adherence to the principle of uniform circular motion, which he realised was a purely theoretical assumption. Ramée's modernism was linked to his religious position, for he was a Calvinist (he was killed in the St Batholomew's Day Massacre of 1572), but this was plainly not the case with Bernadino Telesio (1509–1588) whose criticisms of Aristotle emphasised the role of sense-experience as the true starting-point of natural philosophy. Neither Telesio nor Ramée were active scientists in an experimental sense, but they both fostered empiricism as a crucial foundation of knowledge, and showed that magic and mysticism were not the sole alternatives to scholasticism. One of the most original figures in Renaissance science, Tomasso Campanella (1568–1639), combined empiricism

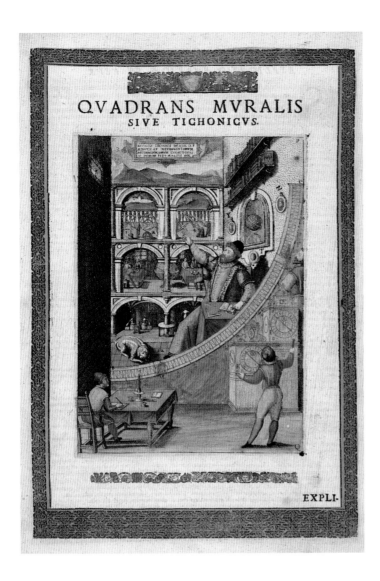

QVADRANS MVRALIS
SIVE TICHONICVS.

EXPLI-

Tycho in his observatory, one of the most celebrated pictures of early modern science. Tycho directs operations while with the huge quadrant one assistant sights a star through a slit in the wall (all Tycho's work was pre-telescopic), a second assistant establishes the time, while a third records everything.

The British Library C.45.h.3.

ISAACVS NEWTONVS.

'A lively and piercing eye': Kneller's portrait of Isaac Newton at the height of his powers in 1689, two years after the publication of *Principia Mathematica*.

Courtesy of the Trustees of the Portsmouth Estate

William Harvey discusses
the problem of mammalian
generation with King
Charles I, using a dissected
deer from the Royal Parks;
the child is the future King
Charles II. Harvey acted
as the king's personal
physician throughout
the Civil War.
Royal College of Physicians,
London

The Orrery, painting by
Joseph Wright of Derby,
c.1763. The orrery, the
mechanical model of the
planetary system, seems
to epitomize the triumph
of the exact sciences in
this period, together with
the philosophical lesson
commonly drawn from
them, that the cosmos is a
vast and intricate machine
pointing unmistakably to
the presence of a designer.
Derby Museum/
Bridgeman Art Library

Testing the cosmic systems. The Tychonic system was preferred by the Church
because it replaced the Earth at the centre of the universe; it clearly outweighs the Copernican
system in this allegorical design published by the Jesuit Giovanni Riccioli in 1651.

The British Library 48.h.3.

with strong mystical tendencies. Wedded to the macrocosm-microcosm doctrine, Campanella pursued the correspondences revealed by magic and astrology, yet was a convinced Copernican, echoing Galileo's distinction between the book of scripture and that of nature. God revealed himself through both, but since the common understanding of mankind could not attain to a scientific knowledge of the world, the Bible presented a figurative account of the creation. This idea was widely repeated by Protestant thinkers in the seventeenth century, indeed it became virtually the official position among Calvinists, but it was not acceptable to Catholic authorities, and the story of Campanella's long years of imprisonment and torture is one of the darkest and most shameful in the history of philosophy. The Catholic Church had become wedded to Aristotelian science in the middle ages, and any challenge to that science was now seen as attack upon Christianity itself. Perhaps ironically, the Neoplatonic tradition that had been so seductive to Renaissance thinkers such as Campanella would soon receive a severe blow in 1614, when the scholar Isaac Casaubon demonstrated that the Hermetic writings were not the authentic records of ancient Egyptian wisdom, but had been composed in the second century AD.

The first systematic defence of science as an independent form of knowledge with its own rationale, appeared more than a century after the private researches of Leonardo da Vinci and the geographical discoveries of the Renaissance, and its source was a strange one. The place of Francis Bacon (1561–1626) in the history of science is full of paradoxes. A lawyer and politician, he was no scientist himself, and seemed unimpressed by the major advances of his day such as Copernicanism. But like his French and Italian contemporaries, Bacon's starting-point was a disenchantment with the scholastic tradition within which he had been taught, a tradition which Bacon argued was dominated by the sterile forms of thought and language which he termed 'idols'. These are to be replaced by knowledge founded on empirical observation, and by inductive argument. This empiricism is not a simple accumulation of facts however, for in the *Novum Organum* of 1620 Bacon sets out rather complex rules for analysing the empirical properties of natural objects in order to discover their essential natures. These rules sound more metaphysical than scientific, and they remind us that Bacon's aim was nothing less than the reform of *knowledge* as a whole. Bacon still has no exclusive word for science as we know it, for physics and metaphysics, mechanics and magic are for him all branches of philosophy, which in turn is one of the four great divisions of learning – the others being history, poetry and theology. Bacon's most explicitly scientific work is the *New Atlantis* of 1627, where, on a Utopian island, teams of scientists collect data and perform experiments, embodying Bacon's ideal of finding 'the knowledge of causes, the secret motions of things and the enlarging of the bounds of the human empire to the effecting of all things possible'. Much of this work was to be directed to practical ends, in medicine, metallurgy, meteorology, food technology and so on. There is no doubt that in Bacon's thought the understanding of nature was to be a prelude to the control of it, to increase wealth and happiness, and in this sense he is as much a prophet of technology as of scientific thought. It is easy to see Bacon as a child of the age of discovery, exploring new realms of thought just as his compatriots were exploring the earth itself, indeed the famous frontispiece to the *Novum Organum* shows a ship voyaging beyond the Pillars of Hercules into untravelled seas. Bacon died before completing the central part of his 'Great Instauration', which was to have set out in detail how his inductive method would have re-classified the world of nature. Whether he would have succeeded in this achievement is uncertain, but, incomplete as it is, Bacon's thought stands at the beginning of scientific discourse in

The new science symbolized by a ship sailing beyond the Pillars of Hercules towards a new world of knowledge: the eloquent frontispiece to Francis Bacon's *Novum organum*, 1620.

The British Library C.54.f.16.

England, and also of the empirical tradition of English philosophy. He refused to agonise over the relation between science and theology, merely stating that our deeper knowledge of the natural world will complement that given by divine revelation; in this too he was part of a long English tradition of 'natural theology'.

From Ramée to Bacon, the theorists of the later Renaissance had probed and undermined the scholastic inheritance of the middle ages. But it was left to Descartes to transform this probing into a total philosophy of scepticism, a questioning of the whole basis of human knowledge. Significantly, René Descartes (1596–1650), a French Catholic, lived most of his adult life in the Netherlands, where his works were published, for, like Italy, France too had claimed her martyrs for the new philosophy. Descartes made profound contributions to philosophy and mathematics, and he produced theories which were long dominant in physics and physiology, yet his thought as a whole seems to pull in two directions: his philosophy is based on the belief in innate ideas, known through what can only be described as rational intuition, while his science claims to be rigorously empirical, and the entire realm of nature is seen as a mechanism. The two aspects of Descartes's thought are linked by the principle of doubt, of radical scepticism concerning all human knowledge, and it is this determined rejection of traditional authority, and the avowed aim of rebuilding knowledge on new foundations, that give Descartes his place in the history of science. Descartes sought to expose the weakness of knowledge on whatever foundation it lay – on authority, on sense experience and on pure reason. Having called all into doubt, he is left with certainty only of his doubting intellect, from which springs his famous dictum *Cogito ergo sum,* 'I think, therefore I am'. But the doubting intellect is only one of several clear and distinct 'innate ideas': others are the existence of God, and of the soul and of the world, ideas which are all of equal status and therefore equally valid. Thus Descartes rebuilds what he has torn down, and easy as it is to criticise his synthesis as an exercise in verbal logic and a new metaphysics, the interest lies in the scepticism, in treating the entire legacy of classical and medieval learning as an insufficient basis for certain knowledge of the world.

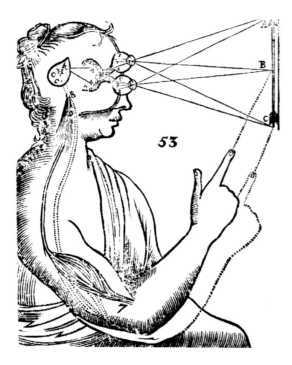

Descartes's physiology: the subject sees an arrow and points to it, and the central processes of cognition and will take place in the pineal gland at the brain's centre; here, Descartes argued, the physical basis of mind was to be found in the gland's subtle vibrations.
The British Library 784.i.1.

In contrast to his intuitionalism when dealing with ideas, Descartes's view of the natural world was thoroughly materialistic, his model for everything in the physical realm being that of a mechanism in which matter acted upon matter. This principle is seen most clearly in his treatment of physiology and of cosmology. The human body in Descartes's view is animated by a fine material substance, the 'animal spirits', which flow through the nervous system, and which in the brain become a 'wind or very subtle flame' which then powers all bodily movement. Descartes is not speaking metaphorically here, for he compares this action to the hydraulic pipes which power the automata and statues 'in the grottoes and fountains in the gardens of our kings'. Descartes conceived an absolute separation of body and mind, or 'rational soul', for the characteristic of all matter is extension, to occupy space, and since mind or soul do neither they must clearly be immaterial. Descartes thus created the celebrated model of man as a ghost in a machine, and the equally celebrated problem of how mind and body can interract. His own answer was that the pineal gland in the brain was the seat of the rational soul, and that its vibrations re-directed the animal spirits and gave rise to complex mental processes such as fear, love or anger. One highly influential result of this theory was that since animals have no souls, they must be mere automata, incapable of consciousness, pleasure or pain.

The mechanical model also shaped Descartes' cosmology, for as a Copernican he was well aware of the problems in physics raised by the idea of free bodies in space. As a convinced mechanist, his solution was to imagine a more ethereal equivalent of the celestial spheres, composed of minute particles gathered into vortices which whirled the planets around in their orbits – a plausible model for

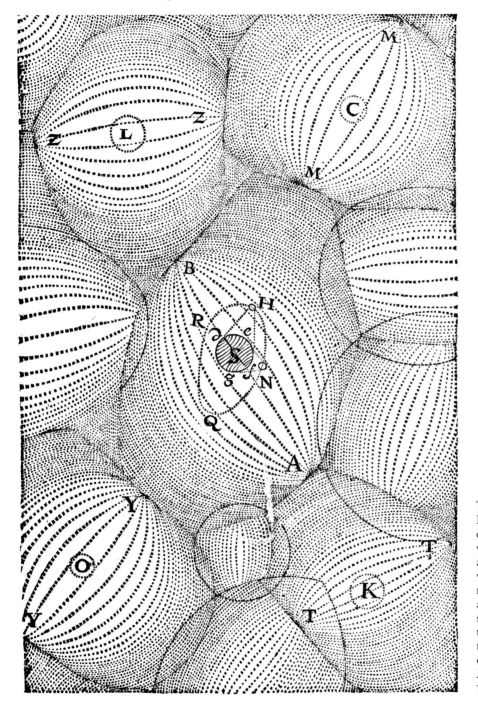

The Cartesian vortices: to Descartes the movements of the heavenly bodies were conceivable only as a physical mechanism in which matter acted upon matter, and he proposed a tenuous ethereal substance which buoyed up the planets and whirled them around in their orbits. From his *Principia Philosophiae*, 1644.
The British Library 1608/737

such invisible vortices was to be found in magnetism. The vortices were ingenious, but were based entirely on the conviction that matter could operate only upon matter, and they are a world away from Kepler's mathematical analysis of planetary motion. In Descartes's view, empty space could not exist, and the universe, instead of being composed of matter and vacuum, was a 'plenum', that is, filled with matter.

The mechanisms of Descartes appealed enormously to emerging science of the seventeenth century, for they represented a real alternative to Renaissance mysticism and magic. But it is difficult to see them fulfilling their empirical programme, indeed despite the long hours which we know Descartes spent on observation of nature, on dissections and experiments, the truth is that he was often wildly wrong in the explanations which he offered for natural processes. To explain what he saw, he invented invisible mechanisms, on the analogy of a machine whose parts were too subtle or minute to be observed. It has been said that he 'tried to put himself at the source of everything, to make himself master of first principles .. so that he could then simply descend to the phenomena of nature as necessary consequences of these principles.' The result is that Cartesian science is not a body of knowledge at all, but a set of attitudes. The legacy which he bequeathed to his many followers was an ambiguous one: on the one hand his rationalism and secularity broke the chains of scholasticism and prepared the way for the Enlightenment in France, but on the other his mechanical philosophy was a blind alley in which his followers wandered for almost a century. Yet, like Bacon in England, he re-defined the structures of knowledge, and gave science a central place within the field of learning. The influence of Cartesianism is visible in many later attempts to survey and systematise human knowledge. In the philosophy of Leibniz for example, for all its metaphysical complexities, what shows through very clearly is the desire to re-state traditional ideas about God, the soul, and creation within the language of physics, so that faith in the order of nature attained a scientific as well as a theological expression.

THE BIOLOGISTS

Long before the end of the sixteenth century, both astronomy and the life sciences had initiated a revolution in the understanding of form, through Copernicus and Vesalius and their successors; but the functioning of these newly-discovered forms was to remain hidden through several further generations of research and thought. Just as the Copernican system required a new physics, so, in revealing the true structure of human anatomy, Vesalius forced a re-examination of traditional, mainly Galenic, theories of physiology. The most famous and far-reaching step was the discovery of the circulation of the blood by William Harvey, announced in his book *De motu cordis et sanguinis* ('On the Motions of the Heart and the Blood') of 1628. By demonstrating a completely new conception of the most fundamental process of the body, Harvey undermined the basis of traditional medicine, and opened the way for a new approach to physiology. Harvey (1578–1657) was not an isolated genius however, for when he travelled to Padua to study at the foremost medical university in Europe, the successors of Vesalius had already made significant progress. Gabriele Falloppio (1523–1562) carried out original research on a wide range of anatomical systems, not only on the reproductive system with which his name is associated through the Fallopian tubes, while Girolamo Fabrici (*c*.1533–1619) first described the venous valves which control the flow of blood back to the heart, and which were to play so important a part in Harvey's theory. Fabrici was also the most important of early embryologists, describing the results of hundreds of observations which he made of hens' eggs at various stages, and raising fundamental questions about the process of generation and the development of the embryo, which physicians would debate for the next two centuries. Perhaps the most intriguing anticipation of Harvey's discovery came in the work of the Spanish scientist and theologian, Michael Servetus (*c.*1511–1553) who clearly knew of the 'lesser circulation' of the blood soon after 1550. In the course of a theological work, Servetus discussed in physical terms the question of the introduction of the divine spirit into the human body. He asserted that this spirit entered the blood with the air in the lungs, and that this revitalised blood was carried by the pulmonary artery to the heart, and thence diffused through the body. Servetus was not concerned to develop the implications of this approach, but he clearly stood on the threshhold of a complete theory of circulation. The work in which he published this idea brought about his death, for in it he repeated his anti-trinitarian views, for which he was condemned to death in Calvin's Geneva, and burned at the stake. Harvey was unaware of Servetus, but learned directly from Fabrici, for he was his pupil and lived for some time in Fabrici's house. Receiving his doctorate from Padua, Harvey returned to practise in London, becoming physician to James I and later to Charles I, whom Harvey attended throughout the Civil War.

From an early date, Harvey developed a belief that life's essential force was immanent in the blood, whose function was to permeate the body with vitality, and he followed Fabrici in considering a drop of blood to be the first feature formed in the embryo. This conviction led him to his extensive investigation into the organic system which the blood requires to do its work. In the physiology inherited by the sixteenth century, the arterial system was believed to carry vital spirit from the lungs through the entire body, a function distinct from that of the veins which distributed nutriment from the liver. It was believed that this blood evaporated in the process of nourishing the body, and was constantly replenished in the liver. The pulse was considered to be a movement in the arteries themselves, unconnected with the heartbeat, since it seemed that the arteries could not dilate simultaneously if the pulse had a mechanical origin elsewhere. Prolonged experiment convinced Harvey that the heart was in fact a pump and was the source of the movement of the blood, dilating the arteries instanteously, 'like an inflated glove'. He realised that the blood expelled from the heart in this way would quickly accumulate to a huge volume, which could not possibly evaporate and be replenished by the liver, an observation supported by the fact that the severing of a major artery leads to death by blood-loss in minutes. There had to be a 'reflux' of blood to the heart. 'I began to consider whether the blood might have a kind of motion, as it were in a circle, and this I afterwards found to be true'. The crucial factor in discovering the slower return flow of the venous blood was the valves, and in a sequence of demonstrations Harvey showed conclusively that they direct the venous blood from the extremities always towards the heart. Thus Harvey concluded that the function of the heart is to cause a constant transmission of blood from the vena cava to the aorta. How exactly the blood from the arterial system was transferred to the venous system Harvey was unable to show, but he postulated the existence of tiny connecting channels, which were indeed confirmed later. More important still, Harvey was unable to explain the *purpose* of this circulation, because the function of the lungs in oxygenating the blood was still unknown. His only recourse was to repeat the traditional doctrine that the heart was the source of the body's heat, and that the blood must there recover the heat it had dispersed to the extremities. It was a further thirty years before Richard Lower (1631–1691) performed a series of experiments which proved that there was no difference between venous blood and arterial blood except contact with the air, from which he correctly deduced the function of the lungs. Despite this incompleteness in his theory, Harvey's discovery stands as the beginning of modern medicine, for it initiated a critical revision of all other Galenic systems, and with the ascendancy of mechanistic science, the circulation of fluids through the body came to be seen as the very essence of life. In addition, Harvey carried out extensive investigations into generation and embryonic growth. Many of his conclusions were erroneous, but he defended the theory of epigenesis – the gradual differentiation of embryo – against those who maintained that the conceptus is perfectly formed and merely grows larger. Harvey was also original in focussing attention on the egg as the first element in mammalian generation, where previously it had been believed that the foetus was composed of menstrual blood. This new approach was summed up in the dictum *ex ovo omnia* ('everything comes from an egg'). In this, Harvey was part of the growing awareness of the 'mystery of life' in terms of secular biology, rather than in terms of divine miracle.

A whole new perspective on this and many other mysteries was opened after 1650 with the invention of the microscope, which, like the telescope in astronomy, revolutionized man's perception of the natural world. The origins of the microscope are more elusive than those of the telescope, but the first publication to include microscopic observations seems to have been a description of bees by Francesco Stelluti, printed in an otherwise unrelated text of 1630. Stelluti (1577–1652) used a simple (i.e. single-lens) instrument made by Galileo which gave magnifications of up to twenty times. He did not follow up this initiative however, and Galileo himself seems to have evinced no great interest in microscopy. In 1660 Marcello Malpighi (1628–1694) was making far more significant use of the simple microscope in Bologna; ten years later Antoni van Leeuwenhoek had constructed similar but superior instruments in Delft; while between these dates Robert Hooke in London had designed and used a compound (i.e. double-lens) microscope. Malpighi's most celebrated microscopic discovery was the blood capillaries predicted by Harvey. 'I could clearly see', he wrote, 'that the blood is divided and flows through tortuous vessels, and that it is not poured into spaces but is always driven through tubules and distributed by

Facing page: Microscope views of the chick embryo by Marcello Malpighi, 1673. Malpighi discovered the capillaries through which arterial and venous blood are linked, and upheld the doctrine of epigenesis – that the embryo becomes structurally differentiated, rather than being preformed and complete from the moment of conception.
The British Library 444.d.23.(3)

Right: Spermatozoa of a dog by Antoni van Leeuwenhoek, c.1673. That on the left was said to be alive, the other dead. Leeuwenhoek's discovery of these living organisms, christened 'animalcules', was of the greatest importance in advancing the understanding of human generation, raising the possibility that they combined in some way with the female egg.
The British Library 462.a.2.(1)

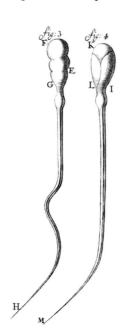

the manifold bendings of the vessels'. He also identified the taste-buds of the tongue, and the red cells in blood, although obviously he was not fully aware of what he was seeing. He wrote a detailed description of the silkworm and another of the development of the chick from the very first hours of incubation. Malpighi was very aware of the new vistas opened by the microscope, and of the enormous task of analysis that awaited biologists. He therefore proposed to confine himself to the simpler life forms in order to build up a picture of fundamental structures, before studying the complexities of human life.

This same approach – the desire to explore the hitherto invisible world of minute organisms – seemed natural too to Leeuwenhoek (1632–1723). Without academic education and speaking only Dutch, Leeuwenhoek was at first isolated from most scientific ideas, and described simply what he saw. Assuming life and motility to be identical, he recognised the tiny moving objects in water as unknown forms of life – microorganisms. His discoveries, including those now classed as protozoa, rotifera and bacteria, were

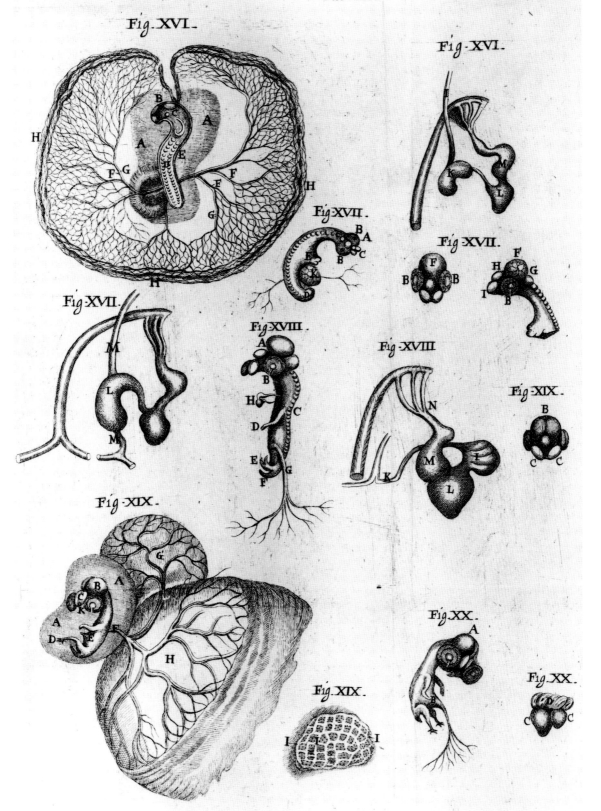

Tab. III.

Fig. XVI.

Fig. XVI.

Fig. XVII.

Fig. XVII.

Fig. XVII.

Fig. XVIII.

Fig. XVIII.

Fig. XIX.

Fig. XIX.

Fig. XX.

Fig. XX.

Fig. XIX.

announced in letters to the Royal Society in London in 1676, where they caused a sensation. Leeuwenhoek was unaware of Malpighi's work, and identified independently the red blood corpuscles and the capillaries. His most important discovery was of sperm cells in semen. These plainly living, moving organisms, which he christened 'animalcules', raised the obvious possibility that they combined in some unexplained way with the female egg to achieve conception. This became an all-important factor in the growing debate on generation between 'ovists' and 'animalculists', which remained one of the keenest biological controversies throughout the eighteenth century. In Aristotelian terms, the male had provided the 'form' and the female the 'substance' of the offspring. Early studies of the development of the embryo however by Fabrici, Harvey and others, emphasized the primacy of the female egg, and were supported by Regnier's de Graaf's description of ovulation in 1672. It was concluded that both oviparous and viviparous reproduction begins with the fertilization of the female egg, and it was further supposed that the egg actually contained the embryo, complete in miniature, which then grew and became differentiated. In this theory the male semen contributed some elusive quality of vital heat or 'vapour'. This view was challenged by the discovery of spermatozoa as living organisms, and by the imaginative pictures of some microscopists, such as Nicolas Hartsoeker, who claimed to find minute human forms – 'homunculi' – in the sperm-heads. To this dispute was added that concerning preformation, for many microscope observers argued that the embryo was fully formed from the earliest moment and merely grew in size, while others believed that they witnessed the gradual differentiation into parts of the embryonic tissue – the process of epigenesis. Both these controversies had a long life and a consensus emerged only slowly in favour of epigenesis, while a true picture of fertilization in which egg and sperm combined, was not achieved until the nineteenth century. A third great biological problem was the phenomenon known as spontaneous generation, where certain life-forms seemed to appear from nothing, the classic example being maggots in cheese or a carcase. It was the Italian physician Francesco Redi (1626–1697) who in the 1660s performed a series of simple experiments proving its impossibility. He sealed meat and cheese in glass vessels while leaving others open to the air, observed insects multiplying on the second group but not on the first, and went on to identify the insect eggs with the microscope. Perhaps as interesting as Redi's experiment is the fact that at this date biological thought was still so fluid that a phenomenon such as spontaneous generation could still be widely believed by scientific men.

The microscope researches of Malpighi and Leeuwenhoek had been communicated piecemeal to the scientific community, but the man who brought microscopy to public attention was Robert Hooke (1635–1702), whose book *Micrographia,* published in 1665, is truly one of the landmarks of science, making it obvious that a new era of scientific data had dawned, and it was in this sense comparable to the impact of Galileo's *Sidereus nuncius.* Many of the book's illustrations were drawn by Christopher Wren, and to their dramatic revelation of insect anatomy, plant tissue and crystal structure, Hooke added novel theories on the functioning of the things which he demonstrated, theories which, whether accurate or not, made it clear that with this new microscopic world, came a huge new task of analysis and interpretation. One of Hooke's most fertile interpretations was to compare the internal divisions in cork to 'cells' in honeycomb; he was not of course seeing cells in the true sense, but when they were identified much later, in the nineteenth century, the term cells was borrowed directly from Hooke.

The most systematic of the early microscopists was Jan Swammerdam (1637–1680), whose main work focussed on the anatomy and life-cycle of

The stem and leaf of a stinging-nettle as shown in Robert Hooke's *Micrographia*, 1665. Hooke's was the most visually striking and influential of the books detailing the revelations of the newly-invented microscope

The British Library G.1524

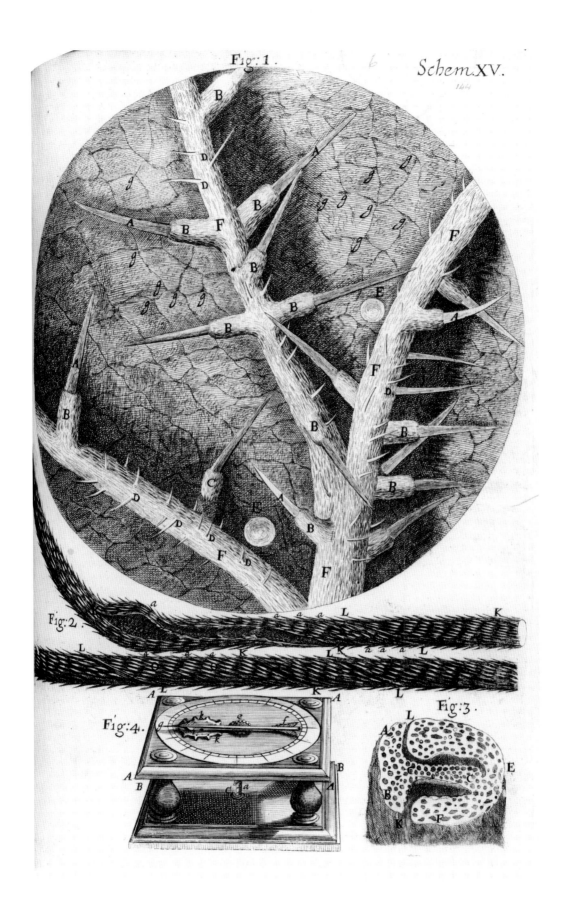

Fig: 2.

Fig: 4.

Fig: 3.

insects. From the time of Aristotle, certain preconceptions about insects had placed them so far below the higher animals that they were considered a form of life barely amenable to systematic study. These were that they arose through spontaneous generation, that they lacked internal anatomy, and that they were subject to swift, inexplicable metamorphoses. Swammerdam's dissections disproved the last two ideas, showing that insects possess systems of circulation, digestion and so on, just as higher animals do. He was able to show that the limbs and wings of butterflies are developing steadily within the caterpillar even before the chrysalis stage, and thus placed the apparently miraculous metamorphosis within the frame of normal growth. Despite his detailed researches, Swammerdam's position in the debate between epigenesis and preformation was ambivalent, and many biologists took his work as support for preformation, arguing for example that the butterfly was always present inside the caterpillar. Religious thinkers gave a radical twist to this belief by proposing that the egg of any species, insect or man, held within it the actual form of the next generation, and that this might be traced back in the case of man to Eve, in whose ovaries were confined the future forms of the entire human race, like a nest of Chinese boxes – indeed the doctrine became known as *emboîtement*. It was conjectured that this might be the mechanism for the transmission of original sin through Eve's descendants, and that when Eve's eggs were used up, the human race would end.

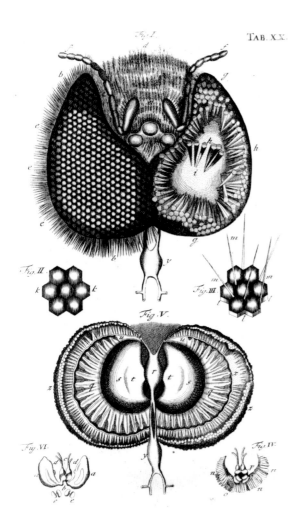

The application of microscopy to the plant world achieved its clearest results in the work of Nehemiah Grew (1641–1712), whose *Anatomy of Plants,* 1682, not only described the structure of stem, root and leaf, but suggested how they functioned. Inevitably many of his ideas were speculative, for example it was considered that the movement of sap within plants might be essentially mechanical, like that of blood in animals, and that there might exist a circulatory or peristaltic motion. Grew's most brilliant insight was to deduce the sexuality of plants: he was wrong in believing that all plants are hermaphrodite, but he correctly identified the stamens as the male organs, and the carpels the female. Grew was not able to substantiate this theory experimentally, but this was achieved conclusively by Rudolf Camerarius (1665–1721) in Tübingen during the 1690s, although his natural assumption that hermaphrodite plants fertilize themselves was later proved to be incorrect. The discovery of plant sexuality was immensely important in establishing a unifying principle of nature that crossed the boundaries of life-forms. It explained the basis of a phenomenon that had been observed and used for thousands of years – the seeding of plants – but whose mechanism had always been concealed.

The task of systematizing and classifying life forms, which had engaged biologists in the sixteenth century, became more complex and bewildering with the

The compound eye of the bee, revealed by the microscope of Jan Swammerdam, *c.*1670. Swammerdam completely disproved the traditional view that insects lacked internal anatomy, and he revealed the distinct stages of their metamorphoses.

The British Library 459.e.4.

advent of the microscope. It was felt, correctly, that throughout the huge diversity of forms and species, a few basic patterns and structures existed, and the search for the secrets of life could not progress until these had been identified. One of the most prescient figures was Edward Tyson (1651–1708), a London physician who pioneered comparative anatomy in theory and in practice. Tyson dissected a porpoise and concluded that it represented a transitional form between fish and land quadrupeds. Exactly in what sense 'transitional' is not clear, for Tyson revived the ancient concept of the great chain of being, but there was no question of a evolutionary time-scale. More important still, he dissected a chimpanzee, and marvelled at its similarity to man, not least in its brain, physically so close to man's, despite the great 'difference of soul'. Empirically, Tyson had arrived at the threshold of the idea that man was an animal; grasping the evolutionary link between man and the other primates would wait almost two centuries more.

It was the science of taxonomy which sought to impose order on the myriads of life forms by distinguishing the primary characteristics of structure and function. Among the most complete attempts to describe a botanical 'system of nature' was that of John Ray (1627–1705), who sought consistent taxonomic principles in the structure and physiology of plants, and fixed on the seed vessel as the most invariable feature. He made the fundamental distinction between monocotyledons and dicotyledons, and described the characteristics of foliage, stem-tissue and life-cycles that are always associated with these types. Although avowing the principle of the fixity of species – that all extant life forms were created by God at the dawn of history – Ray felt compelled to argue that some limited degree of transmutation was possible, and must have occurred, to account for the innumerable varieties of plants that differed from each other in such slight degrees. Ray's magnum opus was his *Historia plantarum,* published in instalments between 1686 and 1704, in which he systematically described more than 5,000 plants, most native to Europe, but attempting also to reduce to order the flood of new plants reported from America and Asia. Even this immense work did not exhaust Ray's energies, for he devised taxonomic principles for birds, fish, land animals and insects too. Ray was an intensely religious man, whose biological research served to reinforce wonder at the richness of creation, and at the principle of design evident in it, indeed he claimed that 'divinity is my profession'.

Science and religion were intimately blended in the mind of one the most original biologists of the seventeenth century, who became the founder of palaeontology, the Dane, Niels Stensen (1638–1686, known often by his Latin name, Nicolaus Steno). Stensen was first a physician, who was responsible for the discovery of the glandular and lymphatic systems, and who recognized that the heart is a muscle, and that its function is to drive the circulatory system, not to re-heat the blood as Harvey and Descartes had supposed. He also rejected Descartes's theories concerning the pineal gland and the 'fluid mechanism' of the nerves. At the age of thirty-five, Stensen turned his back on science, and, having been a devout Lutheran, was ordained a Roman Catholic priest. There is considerable evidence that a religious crisis was precipitated in Stensen's mind by the conflicts between his original researches and the teachings of philosophers such as Descartes, conflicts which led him to doubt the value of human knowledge. His final original work has brought him his most lasting fame, for in 1669 he published *De solido intra solidum,* an early landmark in geological thought. In this work Stensen deduced that certain stones shaped like animal parts were in fact the fossilized remains of long-dead creatures, and he explained the way in which living tissue might be metamorphosed into stone. The specific example he used were the so-called tongue-stones, which were fossil shark's teeth. Stensen introduced the vital concept of stratification, in which layers of the earth may be used as an index of age, and he suggested that mountains are formed by movements in the earth's crust, such movements also explaining the presence of marine fossils on high ground far from the sea. He thus introduced the key concept of chronology into the history of the earth, but he had no true idea of the age of the earth, and he placed this geological activity within the biblical time-frame. There had been several attempts to follow the biblical chronology back to the beginning of history, the most famous being that of Archbishop James Ussher, who dated the creation to 4004 BC, a date widely published and accepted from 1650 onwards.

Seventeenth-century biology from Harvey to Stensen testifies to a revolution in which observation and experiment had become central. In this revolution, the microscope played a key role in expanding the visible world, and the volume and complexity of empirical data made the task of classification ever more difficult and more urgent. However the structures of plants and animals did not reveal themselves in progressively greater detail, because the chromatic aberration of microscope lenses produced rings of colour which obscured the object. Moreover, compared with the structures, the life-functions were still wrapped in mystery, and even so fundamental a process as growth remained incomprehensible. For this reason medical science, as advanced by Harvey and others, had little impact on the crude medical practices of the time. The mechanical philosophy of Descartes was unhelpful here, for it encouraged the view that physiology was a system of forces, pressures or fluids working through a network of vessels. In man, nerves were thought to be hollow capillaries containing subtle fluids, while in the plant world it was conjectured that the penetration of tree roots into soil might be the result of muscular action. In zoology and botany, religion still loomed powerfully in the background, for the interrelation of species could scarcely be understood so long as it was believed that the whole range of living creatures and plants had been created completely and simultaneously at one moment. So observation and experiment became the central activities of the scientist, but intellectual preconceptions still shaped the way he interpreted them. In this respect, Stensen's insight into geological processes was perhaps the most prescient, and his approach would be taken up in the true biological revolution to come in the nineteenth century.

Title-page of the English version of della Porta's *Magiae Naturalis,* a book of experiments and tricks involving heat, magnetism, chemistry. Della Porta's magic was natural in that it could be performed by anyone, and it exploited natural forces: demons and arcane powers were not involved.

The British Library C.48.h.12.

THE PHYSICISTS

In 1589 Giambattista della Porta published a book entitled *Magiae naturalis* ('Natural Magic'), which described among other things methods for hardening steel, games with distorting mirrors, experiments with magnets, a formula for invisible ink, and the properties of lenses. Della Porta was a Neapolitan scholar, the leading figure in an 'Academy of the Secrets of Nature', which met in his house, a man whose beliefs were several times investigated by the Inquisition, and who turned to writing comic dramas to escape their attention. Della Porta's interest in magic had its roots of course in the Hermetic tradition of the Renaissance, but his magic was more than the exercise of a secret power, for he believed that a rational order underlay all the marvels he described. Anyone could perform these feats, for the help of demons was not required. What della Porta was really describing were the *effects* of the forces of heat, light, magnetism, air pressure and so on. He could not yet rationalize or measure these forces, but he stands at the beginning of an experimental approach to them, and those who looked at these

Fire:

Chaos

Ayre:

NATVRAL
MAGICK:
in XX Bookes
by
IOHN BAPTIST PORTA
a Neopolitane :

R.Gaywood fecit Lond:
1658

Art:

Nature:

Earth:

I: BAPT: PORTA:

Water:

forces and their effects from a more rigorous and analytical viewpoint during the following century would lay the foundations of modern physics.

The most single-minded early attempt to investigate one of these forces was the work of William Gilbert on magnetism. The mysterious attractive force present in a certain metal had been noticed and recorded by the Greeks. According to Lucretius the word magnet derived from the province of Magnesia in Thessaly where the metal, likewise called magnetite, was widely to be found. The power of rubbed amber to attract light objects was thought to be another manifestation of the same force: the Greek for amber was *elektron*. This curious power had little practical application until the invention of the magnetic compass, one of the most revolutionary instruments in history. It origin may have been in China, but if so its transmission to Europe has never been clearly traced. The earliest known reference to the compass as an aid to navigation occurs around the year 1200, and it must have become well established by 1269 when the French soldier and scientist, Pierre de Maricourt (also known as Peter Peregrinus) wrote his *Epistola de magnete*, the first full description of the properties of the magnet. Within a few years of this treatise, the first sea-charts of the Mediterranean were being drawn with compass-lines, and the art of navigation took an immense step forward. For centuries however no theory was forthcoming to explain the behaviour of the 'lodestone', although the orientation of the compass was believed to be somehow connected with the north celestial pole.

William Gilbert's pioneering treatise on magnetism *De magnete,* 1600. In this forceful picture, the experimenter creates a magnet by hammering a hot iron bar which is oriented north and south with the earth's magnetic field.

The British Library C.112.f.3.(2)

By the 1580s Copernicanism, Vesalianism and anti-Aristotelianism had penetrated to London and Oxford, as had the vogue for Hermetic mysticism. At Queen Elizabeth's court, these new currents were represented by the bizarre figure of Dr.John Dee, famous as a performing magician, yet a serious mathematician, trainer of the English navigators to the New World, and a recognizable model for Marlowe's Faustus. Exactly contemporary with Dee and physician at the same court was the pioneer of experimental physics, William Gilbert (1544–1603). In the 1580s and 1590s Gilbert was in practice in London, but was also pursuing his own original researches into the nature of magnetism. His results were published in 1600 in his important book *De magnete, magneticisque corporibus, et de magno magnete tellure* – 'On the magnet, magnetic bodies, and the great magnet the earth.' The title reveals Gilbert's central discovery that the earth itself is the focus and cause of all magnetic phenomena. This novel theory was deduced from his experiments with a spherical lodestone, serving as a *terella* or model of the earth, and the way in which it affected magnetized needles placed upon it. Gilbert found that they formed a pattern of meridians, longitude lines, converging at two opposite points corresponding to the earth's poles. Variations from a true north-south alignment he attributed to variations from a perfectly spherical form and to impurities of substance, both of which may be true of the earth itself. Gilbert made an important distinction between magnetism proper and the attractive power of substances such as amber, a force which he was the first to term 'electric', and which he decided was the result of friction. Gilbert's analysis of magnetism was acute, and he was applauded by his contemporaries, including Galileo, while Francis Bacon was to praise his experimental approach. But inevitably he was ill-equipped to account theoretically for what he saw. He argued that every magnet is surrounded by an invisible 'orb of virtue' extending a certain distance and able to affect iron and some other materials. The language is quaint, but describes plainly enough a magnetic field. Yet Gilbert can only rationalize this field as arising from the 'form of the substance' of the earth, implanted in the globe by its creator. All parts of the earth which share this form – lodestone and iron – behave as magnetic bodies. Gilbert cannot define more closely this form, but calls it the primary, radical or astral *anima* of the earth. Gilbert illustrates very pointedly the dilemma of Renaissance science: observation and experiment were spreading, and from them the generalization of certain effects was possible. But although he succeeded in freeing himself from the lure of magic, there was no language of physics or mathematics to carry the explanations further, and extend them into laws.

In this respect the seminal figure was Galileo, who identified different aspects of motion such as acceleration, free fall, impact, and so on, first examining them experimentally, and then finding certain mathematical relationships which governed them. Galileo's work did not form a complete theory of motion, but he plainly banished Aristotle's theory that all movement requires a constantly-acting force. It is possible to see implicit in his work an understanding of inertia, that motion is a state which will continue unless deflected by some outside cause, a perception that became central to Newton. Galileo's influence spread widely through his writings, and in the 1630s Johannes Marci (1595–1667) of Prague carried out important experiments on impact, which led him to deduce the principle of the conservation of momentum, although he was unable to formulate his findings mathematically. Experiment and observation did not automatically command belief in the early seventeenth century, especially if they contradicted established authorities and teachings, for they might be the result of human error, or trickery; when Galileo reported the moons of Jupiter, he was widely disbelieved, and his instruments were said to have produced illusions. Nor, in the absence of scientific equipment, was it clear how to frame experiments and measure their results, and the number of significant experiments in physics remained low for many years after those of William Gilbert.

Around the middle years of the century however, some fundamental experiments on air pressure were carried out in Germany by Otto von Guericke (1602–1686) and in Italy by Evangelista Torricelli (1608–1647). One of the best-known of early scientific pictures shows two teams of horses vainly trying to pull apart two hemispheres enclosing a vacuum. This was von Guericke's first demonstration of the effects of the air-pump, and of the existence of enormous pressure in the air. Descartes had conceived matter as equivalent to space, and had denied the existence of a vacuum: this von Guericke had now disproved, and he was encouraged in his belief that celestial bodies might influence each other across the vacuum of space by magnetic force. Torricelli was an important mathematician, but his name is forever associated with the invention of the barometer. It had long been noticed that suction-pumps, for example those used in mines as described by Agricola, could raise a column of water only to a certain height – approximately nine metres – and no further. Torricelli guessed that that it was the weight of the air which limited the rise in the water, and set out to prove it. He used mercury because its greater weight meant it could be handled in much smaller quantities. He reasoned that, since mercury is 14 times as heavy as water, the atmosphere should support about 0.65 metres of the liquid metal. Filling a tall narrow glass tube, closed at one end, with mercury, he inverted it and placed the open end in in a bowl also containing mercury. The level in the tube immediately settled around 0.65 metres as predicted. Torricelli observed that the level periodically rose and fell slightly, and deduced that the air pressure is not constant, but varies from day to day. It was Pascal who later proved from

The most public and celebrated experiment in early physics: in 1660 in Magdeburg, Otto van Guericke created a vacuum by pumping air out of two joined metal hemispheres, and such was the force of air pressure than two teams of eight horses could not pull the hemispheres apart.

The British Library G.2509.

experiments on the Puy de Dome that the air pressure falls as the height above sea-level increases; from this he inferred that the earth's atmosphere was a finite 'ocean of air' around the planet. (Pascal's physics experiments also revealed the principle of the hydraulic press – that a small weight will balance a much larger one if the surface areas of the water at each end of the system are proportionate.) Torricelli and von Guericke had succeeded in measuring a natural force – the weight of air – whose very existence had been unsuspected just a few years earlier. But what were the implications of these experiments? What other effects might be attributed to such forces, and how could they be described, measured and perhaps related to each other? The answer, as Galileo had prophesied, lay in the development of mathematics.

Classical mathematics consisted of arithmetic for computing quantities and geometry for computing spaces, and both dealt with elements that were static. As seventeenth-century scientists began to explore the world of physical forces, a new mathematics was needed capable of handling elements that were dynamic – variables in time, position, volume and speed. Before mathematics could analyse change and movement, its language must become more powerful and flexible. This process had already begun in the sixteenth century, for example with Francois Viète (1540–1603) who greatly improved algebraic language, employing letters to denote the elements of equations (although not the now-standard x and y which were introduced later). However algebra still employed many Latin words instead of symbols, giving it an unfamilar look. Where Viète wrote:

B5 in A quad – C plano 2 in A + A cub aequatur D solido

we would write:

$5BA^2 - 2CA + A^3 = D.$

Equally important was the innovation of Simon Stevin (1548–1620) who conceived the extension of the Arabic numeral system to numbers less than one, thus introducing decimal fractions, an immense aid to calculation. Stevin made fundamental discoveries in the science of statics, most famously his demonstration of the conditions of equilibrium on inclined planes, in which the effective component of weight is directly proportional to the angle of the plane. This principle led Stevin to his original statement of the parallelogram of forces, where two forces are equivalent to one along a line of intersection. This rule, and the associated concept of zero forces in equilibrium, have become the cornerstones of statics. Stevin provides a clear example of a new mathematical approach to physical effects which had perhaps been observed and used, but never analysed. Similarly Willebrord Snell (1580–1626) succeeded in formulating the law of refraction which bears his name, setting out the constant ratio of refraction as light passes from one medium to another, air to water being the classic example; from this it followed that different substances all possessed their own refractive index.

Simon Stevin's picture of the inclined plane and the equilibrium of forces, 1586. The downward pull of the four chain links on the left exactly balances that of the two on the right, because the angle of the latter is twice that of the former. This led Stevin to the parallelogram of forces, and the foundations of statics.
The British Library C.112.d.2.(1-3)

Calculation was enormously simplified with the publication in 1614 of the first logarithm tables by John Napier (1550–1617), where addition and subtraction of the logarithm takes the place of multiplication and division. The logarithm of a number is defined as the number of times a base number must be multiplied by itself to raise it to that number. The process is based on pairing an arithmetical sequence with a geometric sequence. Napier's logarithms were somewhat simplified by Henry Briggs (1561–1630) who took 10 as his base. For example:

Arithmetical sequence:	1	2	3	4	5
Geometric sequence :	10	100	1000	10,000	100,000

To multiply 100×1000, add $2 + 3 = 5$; 10 raised to the power of 5 is 100,000. Tables of logarithmic functions were essential tools of the mathematician until the invention of the modern calculator. A related aid was the linear slide rule, invented by William Oughtred (1575–1660), which, when mastered, also functioned like the calculator.

Descartes made his contribution to mathematics in 1637 in which he applied algebraic methods to a geometric figure. He extended the theory of co-ordinates, and of lines or curves as functions of a point moving in relation to fixed axes, the concept underlying the fundamental mathematical model of variation, the graph. Descartes simplified the language of algebra by using *a*, *b* and *c* to represent known quantities, and *x*, *y* and *z* for unknowns. He also devised the modern exponential notation x^2, x^3, and so on. Analytic geometry flourished particularly in France among Descartes's contemporaries, Pascal and Fermat.

In the careers of Gilbert and von Guericke we see experiment leading to hypothesis. In Stevin and Torricelli we see experiment combined with measurement, leading to the formation of mathematical laws. Evident in the work of all these men is the conviction that mathematics could become a system able to describe the structures of the physical world as language could not. Whether this was because mathematics corresponded to the actual structure of the universe, or whether mathematics was merely a convenient set of symbols, was a metaphysical question familiar since the age of Pythagoras. But what was clear by the later seventeenth century was that concepts alone – acceleration, attraction, force, vacuum, and so on – were no longer enough, and that research in physics was passing beyond language into the realm of mathematics.

This movement reached its apogee in the career of Isaac Newton (1642–1727), who built a new vision of the universe upon mathematics. Newton's achievement marked the culmination of the scientific revolution by providing the new basis in physics demanded by the Copernican system and by the abolition of the celestial spheres. Newton as a man was solitary, severe, sometimes fanatical, and highly vulnerable, but these qualities plainly sprang from his genius: he was a man set apart from other men, and he knew it. His childhood was bleak, and he seems to have shown no interest in natural philosophy or mathematics until his undergraduate years in Cambridge. Then he discovered the mathematics of Descartes, and in a few short years and without tuition, he mastered the most advanced analytical algebra, and then surpassed it by devising the language of calculus. This he did in private, without publishing his results, so that by the age of twenty-three, Newton had become the most original mathematician in Europe, although no other scholar was even aware of his existence. In 1665 the university was closed by the plague, and for almost two years Newton lived in rural solitude in Lincolnshire, a period during which some of his fundamental perceptions on motion and on light took shape. The story of the fall of the apple in the orchard was set down much later, but it may well be a true one, and if so it must date from this period.

Pondering on circular motion and the fundamental problem of the earth's orbital velocity, Newton devised a method of quantifying the centrifugal force which, according to commonsense, should hurl all matter from the earth's surface. Huygens had investigated centrifugal force in the 1650s, indeed had invented the term, but his results remained unpublished. For a rotating body as massive as the earth, Newton calculated that this force was a huge one; yet the simple fall of an apple from a tree to the earth showed that it was counteracted by one still more powerful. How could this other force be quantified? Taking a long pendulum, Newton timed its fall from rest as 200 inches per second: this was the acceleration of gravity at the earth's surface. Using very approximate values for the size of the earth, Newton calculated that the power of centrifugal force to this other force was only 1:300. To this other force he would later give the name gravity, the Latin word meaning simply 'weight'. But if gravity extended upwards, as it plainly did, to the tops of trees or towers or mountains, why not further? Suppose that the Moon, also in circular motion, and also 'endeavouring to recede from the earth' with centrifugal force, were subject also to gravity, how could it be quantified? By calculating the centrifugal force operating on the

Moon, and comparing it with the gravity at the surface of the earth, Newton found that gravity is some 4,000 times stronger. But Newton had now seen that Kepler's third law could also be seen as a statement about centrifugal force, and that the 'endeavour to recede' from the centre will be as the squares of the distances from it. By this reckoning, using 60 earth radii as the earth-moon distance, the ratio should be 1:3600, which in Newton's words 'answered pretty nearly' to his figure, but not closely enough to be regarded as a law of physics. Newton laid aside this problem, and almost twenty years of mature thought were to pass before the full inverse-square law of gravitation was reached. But once again in private study, this solitary 24 year-old had entered realms of thought unknown to any scientist in Europe.

Newton now turned his attention to light and performed experiments with prisms which would lead to his revolutionary theory of colour. All scientist and philosophers had spoken of colour as something added to white light or as a modification of it. Observing a thread joined of red and blue cotton through a prism against a black background, Newton recognized that that one straight line of thread is never seen, but that the red is always higher, because of the unequal refraction of the two colours. White light could not be brought to an exact point focus, because the focal length of each colour was different. Light was clearly heterogeneous, and colours resulted from the splitting of light into its components. To confirm this, the conclusive step was to place a second prism which re-combined the colours split by the first back into white light. But when any single colour, isolated by slits in a board, was directed through the second prism, it remained intact. Thus it is white light which is derivative or secondary, and not the colours. Newton now understood that the problems of chromatic aberration experienced with lenses in telescopes and microscopes were inherent in the process of refraction, and he set himself to build a reflecting telescope which overcame the difficulty, and which caused a sensation when it was demonstrated to members of the Royal Society. As to the mysterious nature of light, Newton adhered to the common belief in its physical reality, that it consisted of minute corpuscles. This belief seemed to be proved in a striking way when the Danish astronomer Ole Rømer deduced in 1676 that light travelled at a finite speed. Rømer reached this conclusion by observing that eclipses of Jupiter's moons were slightly advanced or delayed from their predicted times as Jupiter moved closer to the earth then away from it. He calculated the speed as 140,000 miles per second, some 25% lower than the true value.

Newton's demonstration of his telescope in 1671 marked his emergence into public life, for although he had returned to Cambridge after the plague years and had been appointed to the chair of mathematics, he was still quite unknown outside the university. By publishing papers on his researches and corresponding with other scientists, Newton was drawn into a succession of personal confrontations, in which criticism drove him to an irrational fury and into self-imposed seclusion. Yet it was in response to questioning and probing from his contemporaries that Newton was moved to take up once more the problem of planetary motion, and to produce his great synthesis of astronomy and physics. While Newton had isolated himself in Cambridge, the most distinguished scientists in London – Halley, Wren and Hooke – were meeting to discuss, among other things, planetary motion, and they too made an intuitive connection between Kepler's third law and the attraction between the bodies of the solar system. The story has become famous that Edmond Halley, an outstanding intellect and a genial, diplomatic man, with whom even Newton never quarrelled, travelled to Cambridge in the summer of 1684 to ask Newton what the path of a planet

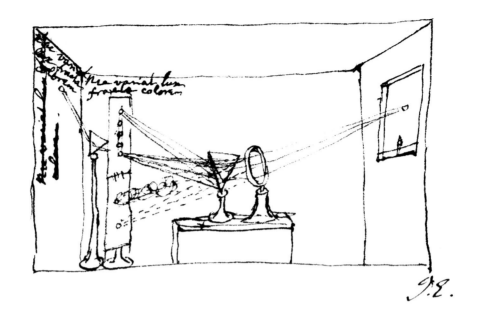

would be if the force of attraction towards the Sun were reciprocal to the square of its distance from it. Newton replied at once that it would be an ellipse, and Halley, 'struck with joy and amazement', asked how he knew it, to which Newton answered that he had calculated it, and promised to send on his proof. In the event, this proof was two years in the writing, for it grew to be *Philosophiae Naturalis Principia Mathematica* – 'Mathematical Principles of Natural Philosophy', which, when published in 1687, offered a new physical theory of the cosmos. Although Newton was as concerned as Descartes with physical explanation, his mechanics were fundamentally different, for they were not hypotheses or chains of logic, but were precise, quantitative descriptions of the motions of real bodies, expressed mathematically. First Newton addressed terrestrial motion, and defined some original and fundamental terms: force, mass, inertia, centripetal force and so on, which created a language of dynamics as the essential prelude to his analysis. From these elements he then formulated the three classical laws of motion: that a body remains in motion unless compelled to change its state by an outside force; that this change of motion is directly proportional to the force acting on it; that to every action there is an equal and opposite reaction. With these laws, Newton used mass and velocity to calculate the force necessary to divert a body from linear to circular or elliptical motion.

'I have laid down', he wrote, 'the principles of philosophy; principles not philosophical but mathematical. … These principles are the laws and conditions of certain motions, and powers or forces. … It remains that, from the same principles, I now demonstrate the frame of the System of the World'. Turning to celestial physics, Newton proceeded to abolish forever the ancient imagined distinction between the physics of the earth and that of the heavens, for the dynamics which he had defined applied to the planets too. With Kepler's third law before him as a model, Newton found that the force holding the planets in their orbits (now termed a centripetal as opposed to a centrifugal force) decreases as the planets' distance from the sun. Newton had obtained from Flamsteed, the Astronomer Royal at Greenwich, precise data on the positions of the Moon and planets, and all the evidence supported his intuition that the same principle governed motions throughout the solar system. He was at last able to generalize the law of universal gravitation, that every particle of matter in the universe attracts every other particle with a force directly proportional to their combined mass and inversely proportional to the distances between them. All historical theories of celestial motion – Aristotle's constant force, Descartes's vortices – were obsolete, replaced by an inertial system sustained by action at a distance. The vortices were shown to be impossible because they were unsustainable; their motions had no energy source, and would dissipate themselves.

Newton's most striking quality was his intuitive grasp of physics, especially when venturing into virgin territory where no one could guide him. He seemed able to penetrate to the heart of complex problems, typically involving geometry and movement with several variables, and to seize unerringly on the answer; but having done this, he would calculate the mathematical proofs with the utmost rigour. As Beethoven laboured tirelessly to give the most perfected forms to the musical themes which his inspiration gave him, so the muse of physics seems to have granted to Newton a creative brilliance, which he repaid with days and nights of dedicated computational labour. Many scientific discoveries have a degree of inevitability – had Galileo not seen the moons of Jupiter for example, another observer would eventually have done so; but an intellectual synthesis such as Newton's was unique in its time and possibly unrepeatable, and the course of scientific history would have been quite different without him. His

A page from Newton's notebook which shows his analysis of white light, refracted through a prism to reveal the constituent colours. When each colour is passed singly through a second prism, there is no further refraction, and it remains intact. Not shown in this sketch is the recombination of colours through a second prism into white light. The text below is unrelated.

By permission of the Warden and Fellows of New College, Oxford masterpiece was greeted with wonder, and was immediately recognized as changing man's intellectual landscape. But the wonder was not unmixed with bafflement, for Newton's principle of action at a distance was far more mysterious than any mechanism. The Cartesian school of philosophy, the great figures of Leibniz and Huygens for example, accused Newton of reviving the occult forces which the mechanical philosophy had sought to banish, and they could not understand why Newton failed to theorize about the mechanism of gravity. This reaction underlines the fact that, like all the greatest scientific discoveries, Newton's handling of gravitation was an epistemological revolution. This is the significance of the title of Newton's book: a science of mechanisms such as Descartes's proceeded by visualizing structures and chains of connection, while action at a distance and the effects of gravity demanded to be analysed mathematically, and no hypothesis is offered as to the origin and nature of these great forces. Thus one set of questions and problems was answered by means of a new language, a new perspective and a new standard of explanation. Yet, like a horizon that is always receding, fresh problems immediately arose, which even the new language was unable to answer. Mathematics had given Newton his proofs that gravity was universal; but those mathematics could not explain what gravity actually *was*, and Newton himself could describe but never define it, nor how the solar system had reached its perfect equilibrium of mass and distances. Today as in the seventeenth century, gravity is known by its effects, but not as a physical entity.

Newton was the towering intellectual figure of his age, but he was of course part of a rich scientific revival in England, which found its expression in the formation of the Royal Society. Edmond Halley (1656–1742) who played so important a part in the publication of Newton's *Principia,* had travelled to the island of St Helena in order to map the stars of the southern hemisphere, and he later compiled the first sea-charts showing magnetic variation, which became an invaluable aid for navigators. His name is always linked with the comet of 1682, whose path he analysed so closely that he was able to identify it with comets recorded in 1531 and 1607, and to predict its return in 1758. This event was seen as a triumph of Newtonian astronomy, and as putting an end to the superstitious fear with which comets were regarded. Halley was the first to demonstrate that the stars were not fixed, but had motions of their own, for the discrepancies in star catalogues from the time of Ptolemy onwards were too great to be explained in any other way. Halley could not quantify these movements, but they were verified experimentally in the nineteenth century.

After Newton, the scientist from this Royal Society group who had the greatest significance for the future was Robert Boyle (1627–1691), an original experimenter in physics and the proponent of a new, rational approach to chemistry. Chemistry was actively pursued in the seventeenth century, so that anecdotal and experimental data continued to accumulate, but there was no general theory, no conceptual framework within which to integrate this knowledge. Its twin roots in alchemy and mining-metallurgy encouraged on the one hand a degree of secrecy, and on the other elaborate metaphysical systems personal to each practitioner. The case of Jan Baptista van Helmont (1579–1644) is typical, for he made genuine experimental advances, while wrapping his discoveries in a mystical language reminiscent of Paracelsus. Van Helmont recognized the existence of gases distinct from atmospheric air, indeed he invented the word gas (the word is derived from 'chaos') after experiments with burning and fermenting. He reduced the classical four elements to two – air and water – and considered water to be the fundamental constituent of matter. This he proved by the celebrated experiment of feeding a sapling willow tree for five years on water alone, and measuring its weight gain, which was 164 pounds, while the soil appeared undiminished. This conclusion shows how misleading an experiment might be without an adequate conceptual framework. Moreover van Helmont was still a prey to the alchemist's temptation to see within all chemical processes the workings of spiritual principles, for example when a substance was converted by fire into gas, it was said to have shed its earthly shape and reverted to its pure essence, its spiritualized form.

It was precisely this kind of language that Boyle wished to escape from, replacing qualitative description with mechanical relationships. He had begun with physics experiments, building, with Richard Hooke's help, much improved air-pumps with which to test the properties of the vacuum. He demonstrated that in a vacuum respiration was impossible, that sound did not carry, that light objects fell at the same rate as heavy ones, and that most substances, even sulphur, would not burn. These were important findings, but Boyle missed the crucial fact that only a part of the air was involved in respiration and combustion. This was proved just a few years later by John Mayow (1641–1679), who burned candles in closed containers until they went out, but found that most of the air still remained. Boyle demonstrated the law that bears his name by compressing air in a U-shaped tube of mercury, having one end closed. With air trapped in the closed end, he added more mercury to the other, until that column stood at 59 inches, or two atmospheres. Boyle reasoned that the trapped air had been compressed to half its original volume, and that this was reversible. Thus the volume of a gas or air varies inversely as the pressure upon it, and a gas will expand through the available space; in addition to weight, air clearly had elasticity, which Boyle called 'spring'. In the significantly-titled *Sceptical Chymist* of 1661, Boyle attacked the obscurity of the alchemists, and the doctrine of the four classical elements or the three Paracelsian principles, arguing that none of them had ever been isolated from mixed substances. Instead he stated:

> 'I now mean by elements certain primitive and simple bodies, which not being made of any other bodies or of one another, are the ingredients of which all those called perfectly mixed bodies are immediately compounded and into which they are ultimately resolved.'

This sounds very close to the modern concept of an element, and Boyle did indeed embrace a form of atomism to explain the chemical reactions, but it has as yet no quantitative basis. Instead Boyle infers that the basic particles of matter – 'corpuscles' – have different shapes and qualities: the corpuscles of acids are spiky, which explains their sharp taste and their ability to break up other substances. Oils on the other hand have smooth corpuscles, while those of salts and metals are geometrically-shaped, so that they can bind into their respective crystals. This corpuscular theory was very influential, especially when applied in a mechanistic, Cartesian manner, for it was taken to explain all the secondary qualities of matter – weight, colour, smell and so on. If the qualities of all corpuscles could be known, then the operations of all material substances could be understood. As John Locke wrote 'The dissolving of silver in *aqua-fortis* and gold in *aqua-regis* and not vice versa, would then perhaps be no more difficult to know than it is to a smith to understand why the turning of one key will open a lock and not the turning of another'. This emphasis on the physical qualities of atomic particles would prove to be a blind alley, but Boyle had given a new direction to chemistry, lifting it to some extent from the alchemists' dreams into the daylight of rational study. Yet chemistry was to prove resistant to rationalization for many more years: alchemy still flourished even as the great academies of the new science were being founded. Boyle did not regard the transmutation of metals as impossible in principle, and even Newton, despite years of intense study, was unable to create a sound theoretical framework for chemistry.

SCIENCE, CULTURE AND BELIEF

It was not only the content of seventeenth-century science that was new, for the social context and the status of scientific thought were also transformed. This happened not through the universities, which remained officially wedded to Aristotelian science, but through private groups of scholars who corresponded with each other and, less commonly, met to exchange ideas, thus forming 'invisible colleges'. The earliest that we know of were in Italy, the 'Academy of the Secrets of Nature', led by della Porta in the 1580s, and the 'Academy of the Lynxes' (so-called because of the lynx's celebrated clear-sightedness) founded in Rome in 1603, whose most famous member was Galileo. The Lynxes dissolved in 1630 with the death of its leader, Prince Federico Cesi, thus avoiding what would surely have been a fatal conflict with the Church following the Galileo trial. In Paris, Marin Mersenne, in addition to hosting a salon in the 1630s, where figures such as Descartes, Pascal and van Helmont met, kept up a correspondence with natural philosophers all over Europe, circulating ideas and news of discoveries. Mersenne's group, termed the Academia Pariensis, would form the nucleus of the Académie Royale des Sciences, founded in 1666, just six years after the formation of the Royal Society in London. Before this, London's Gresham College had acted since 1598 as a focus for the new science, for alongside the traditional disciplines of law and divinity, lectures were given on physics and mathematics as applied to topics such as navigation, surveying and engineering. After 1660 the Royal Society became the meeting-place of the most distinguished group of scientists ever seen in England – Wren, Halley, Boyle, Hooke, and Newton, as well as attracting foreign members such as Huygens, Leibniz and Leeuwenhoek. For the first fifteen years of the Society's existence, the relatively humble bookseller and mathematician, John Collins, fulfilled a function similar to Mersenne's in France, of linking by correspondence scientists in England, the Netherlands, Germany and Italy, becoming the semi-official 'intelligencer' of the Society. Other important academies would soon be founded in the Netherlands, Prussia and Sweden, and the journals in which they published scientific papers gradually supplanted private correspondence as the means of exchanging ideas and information. These publications – such as the *Philosophical Transactions* of the Royal Society – were most important in building a body of knowledge and a consensus of scientific thought, where one man's work could be based on or react against another's. Few of the great figures in seventeenth century science were university teachers, but lived by private means or private patronage. If they did hold university chairs, the new science played little or no part in their teaching; Newton for example did not teach calculus or gravity theory to Cambridge students.

The formation of these academies demonstrates that the ancient study of natural philosophy had enormously enlarged its scope as a body of knowledge, had become a profession, and was seeking a means of self-expression. This self-consciousness can be seen in the frequent priority disputes in which Galileo, Newton and others became involved: a new intellectual realm was plainly taking shape, and its leading figures were anxious to define their role and authority in it. As this century of scientific revolution drew to a close, the most difficult retrospective question of all is to decide how far the status of scientific thought had changed: did people understand science as having somehow re-interpreted their world? Did the language of mathematics, physics, experiment, and measurement force a re-evaluation of traditional beliefs about the way nature worked? Had science acquired an authority of its own which could rival older and traditional beliefs?

It has been argued that science has replaced Christianity as the dominant force in European civilization, and it would be natural to discover the beginnings of this process in the seventeenth century; but this view cannot be easily sustained against a number of powerful counterbalancing facts. In the first place, the seventeenth century was, perhaps surprisingly, not a great age of technical innovation. The telescope and the microscope were revolutionary instruments, but their force can hardly have been felt in the world at large. The most widely-admired invention was probably the pendulum clock devised by Christian Huygens (1639–1696). Galileo had drawn attention to the striking mechanical properties of the pendulum, and had suggested a possible application to clocks, but it was Huygens who made the transition from theory to practice. With this exception, technological change in areas like power, transport or engineering lay still in the future, and the general impact of the new science was therefore not great. Secondly, the supreme authority in all intellectual matters was the Christian religion, and any new system of knowledge and belief would have to measure itself against religious teachings. Yet it is impossible to trace at this stage a power-struggle between science and religious belief. The essence of Protestantism was supposedly the individual conscience and judgement, the personal experience of the divine. Yet the scientific revolution was nurtured in Catholic Italy and France, and the conflict between science and the Church centred on astronomy alone, with mathematics, biology and chemistry remaining theologically neutral. Cartesianism was integrated with religion so successfully by Catholic theologians such as Malebranche, that it became for a time a new orthodoxy as strong as the old Aristotelianism. It is true that science took root very strongly in the northern, Protestant countries – England, the Netherlands and Sweden – yet France and Italy never ceased to produce scientists of the highest importance, and France sponsored state science as England never did at this time, in fields such as geodesy, mapmaking, transport and manufacturing. Thirdly, the writings of individual scientists themselves overwhelmingly seek to place their work within a religious world-view. Galileo himself always remained a devout Catholic, repeatedly seeking permission to leave his house-arrest to attend mass. Kepler was temperamentally a mystic, seeking always the finger of the creator in the creation. It comes as a surprise to learn that John Napier's greatest energies were devoted to religious controversy: to the suppression of 'papists, atheists and neutrals', to the design of war-machines to defeat the enemies of religion, and to calculations concerning the end of the world. Boyle promoted missionary work abroad, and published in 1690 *The Christian Virtuoso,* written to advance the view that the scientific study of nature was a religious duty. The clearest case of all is Newton himself, whose religious fervour was as deep as it was unorthodox. He spent long years immersed not in mathematics and optics but in theological and biblical studies, and produced works on the trinity and biblical chronology which were published many years after his death. He was a passionate anti-trinitarian, and despised the establish churches as a fraud. His personal religion was highly rationalized, expressed in his favourite quotation from the Wisdom of Solomon, that God created everything 'by number, weight and measure'. Nature, he believed, 'does nothing in vain', but rules through 'the order and beauty which we see in the world'.

FIG.I.

FIG.II.

FIG.IV.

FIG. III.

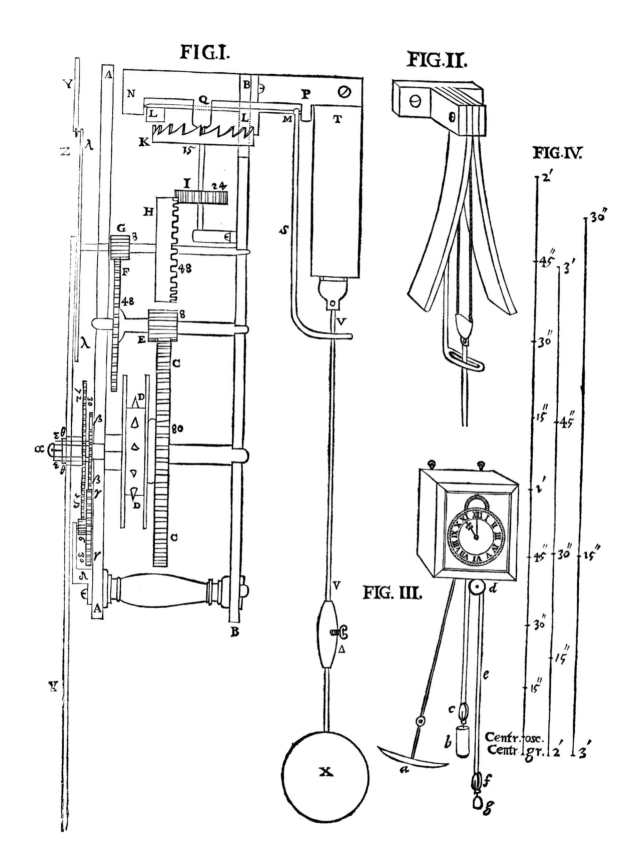

The central concept which enabled these and other scientists to integrate their discoveries within a religious framework, was that of the 'laws of nature'. In a general sense this was not a new concept in the west, having roots in medieval naturalism, the belief that God ruled the world through natural mechanisms and natural chains of causation. What was new in the Newtonian age was the conviction that these mechanisms were expressible in the precise mathematical language of force and velocity, ratio and proportion. The doctrine of the two books – Nature and the Bible – was likewise an ancient one, but the book of nature was no longer accessible to reason in a general sense, but must be analysed mathematically into circles, ellipses, triangles, formulae and relationships. Newton was as convinced as the mystic Kepler had been, that only this language would reveal the underlying harmony built into the universe by its creator. This harmony was expressed in the 'laws of nature', a term which perhaps consciously echoed the divine laws of the book of Genesis, thus underlying the parallel between science and religion. The laws of motion, of optics, of pressure, of gravity, all expressible as mathematical formulae, were the great discoveries of seventeenth-century science, and they generated a new vogue for rational theology in which design and order in the universe were seen as demonstrating beyond all doubt the mind of a supreme creator. The pendulum clock developed by Huygens became the model of this sense of design in nature, so that the universe was conceived as a vast and complex mechanism that was unthinkable without a rational God. This mechanism was rather simplistically envisaged by the Cartesians, more subtly by the Newtonians, but the more subtle the mechanism, the more miraculous was the creator's power, so that this vein of natural theology could be richly mined well into the future. The model of the experiment began to be copied by theological writers: people were invited to consider dispassionately the complexity and order of the world, and to deduce its source – only one answer being possible.

In shaping this image of the cosmic mechanism, the central sciences were obviously astronomy and physics, and these were the sciences which had made such gigantic strides, indeed the scientific revolution defines itself essentially in their terms. The biological sciences still needed far more data before they could progress, while chemistry and geology had scarcely emerged from the darkness. There was as yet no horizon of knowledge in these latter sciences, no agreed consensus on even fundamental matters. Above all it cannot really be argued that knowledge had become truly *secularized* in this period, for the authority which science had acquired was used by its leading practitioners to reinforce religion. It is true that science had developed a language of its own, in some cases a very powerful and technical language, but the context in which that language was exercised was still that of a religious world view, and would remain so, unchallenged, for many more years. Mathematics was a tool which could analyse and measure the world, but could offer no explanations of it. Knowledge was secularized only to the extent that if one wished to know what was happening in an eclipse, how the heart functioned, how to measure planetary parallax or how a pulley achieved its effect, one would not search the Bible or ask a priest; but then obviously no one ever had sought answers to those questions put in those terms: it was the questions and the language that were new, for science had not yet changed the world.

Finally, is it possible to speculate why the scientific revolution occured when it did? And why it occured only in European and not in Islamic or Chinese thought, which had so recently been more advanced than that of Europe? The conditions had been created during the late fifteenth century, with a rediscovery of the world that was both geographical and aesthetic. It was only later, in the seventeenth century, that this rediscovery was given an intellectual form. The old

Christian Huygens's pendulum clock of 1673. Galileo had noticed the isochronic swing of a pendulum, and had suggested its potential for time-keeping, but it was Huygens who perfected the design, and it proved far superior to the weight-driven clock.
The British Library 48.f.16.

Science *à la mode*: a fashionable couple discuss planetary astronomy. By the late 17th century, no cultivated person could remain ignorant of the discoveries made with the telescope and the microscope. Notice the satellites of Jupiter and Saturn, and the representation of day and night on all the planets. From Fontenelle's *Entretiens sur la pluralité des mondes*, 1686.

The British Library 1479.a.6.

world, the classical world, was exploded by the navigators of the age of exploration, and in the same way the ancient forms of philosophy and knowledge must
be tested, reformed or replaced. Scholasticism, abstract theories of quality and
purpose, were no longer sufficient, and the new science sought to formulate
testable propositions about the real world. Bacon's image of the ship sailing
beyond the confines of the old world perfectly expresses this new ambition. In the
long interval between the re-drawing of the world map in the 1490s and the
emergence of the new science a century or more later, the essential tool was being
developed – the language of mathematics, without which quantitative laws could
not be expressed, or even conceived. Coincidentally, the demands of navigation
acted as a clear stimulant to the pursuit of mathematics, giving it very tangible
goals, goals that were political and financial as well as intellectual. Thus, if it is
possible to identify a seminal influence in the growth of western science, it was
surely the new geographical discoveries of the years 1490–1520. Their significance took some time to absorb, but their impact on the European mind proved
decisive in both intellectual and technical terms, while the other advanced cultures of the fifteenth century, Islam and China, experienced nothing remotely
comparable. The subsequent growth in European science and technology would
be determined by a view of nature and of man's relationship to it that was first
articulated in the century of scientific innovation between Tycho and Newton.

Chapter Six

EIGHTEENTH-CENTURY INTERLUDE

&❧

' I am a good encyclopedist,
I know what's true and what's false,
I am Diderot on the track:
I know everything and believe nothing.'

Denis Diderot

THE VINDICATION OF NEWTONIANISM

Newton was born in 1642, which was the year of Galileo's death, and the lifetime of these two men defines for us the period of the scientific revolution. By the time of Newton's death in 1727, science had become a major force in western intellectual life, exerting a powerful influence on philosophy and theology. The education of a gentleman still centred on the classics and mathematics, but no cultivated man could remain ignorant, in general terms, of Copernican astronomy, the theory of gravity, or of concepts such as vacuum, chemical reactions, magnetic force, and so on. Science was in fashion, with lectures and experiments providing a form of entertainment for the salons. This is the familiar image of eighteenth-century science: elegant figures in silk and lace gazing impassively at air-pumps, electrical machines, and dissected animals.

Yet in terms of intellectual innovation, the eighteenth century unfolded as a prolonged anti-climax after the scientific revolution. There was no second Newton in physics, no second Galileo in astronomy, no discoveries in biology to compare with those of Harvey, Leeuwenhoek or Swammerdam. The eighteenth century was an age of science in the sense that an awareness of science spread widely among educated people, but not in the sense of major discoveries or innovations. This was the age of the botanical garden, the scientific society, and the 'cabinet' of natural curiosities; beyond this, science was scarcely visible. There was indeed a technological revolution in the making in England, in steam power and in iron and textiles, but it had a craft basis, and its pioneers had no contact with the intellectual world. The new machines were built in mining regions, hidden from polite society, and as yet had little impact on people's lives; such a pioneering book on economics as Adam Smith's *Wealth of Nations*, published in 1776, made no mention of steam engines, which had by then existed for sixty years. There were no great reforms in education to reflect the scientific revolution, for the universities remained wedded to a virtually medieval syllabus of classical literature and mathematics, while scientific men met in their private academies. Science in this period was characterized by advances in three areas: there were far-reaching developments in mathematics, many of which served to vindicate Newtonian physics; there was a rapid accumulation of data, notably in zoology

Memorial painting to Newton by Giovanni Pittoni, 1730. By the mid-eighteenth century, veneration of Newton reached unprecedented levels, and not only among scientists, for the light he had thrown on the structure of the universe. Here the vast urn containing the remains of the hero is surrounded by symbols of mathematical science.
Fitzwilliam Museum, Cambridge

and botany, together with a universal desire to measure and classify natural phenomena; and there was a philosophical enquiry into Nature and its laws, and the
role of God and man in the new science, an enquiry which came to very different
conclusions in England from those in France. Only towards the end of the century did the stirrings of a second scientific revolution begin to appear, with the
new sciences of electricity, of empirical chemistry, and with the revolution in
industrial machines.

Many European philosophers found Newton's celestial physics difficult to
accept, and Descartes's mechanistic science remained for them a more attractive
model than action at a distance. Newton's dispute with Leibniz over the priority
in discovering calculus tended to reinforce the tension between English and continental scientists. Newton discovered the technique of calculus first, but Leibniz
published first, and the notation which Leibniz devised was clearer, and proved to
be more attractive to other mathematicians than Newton's own. Strangely therefore, the history of mathematics, in purely technical terms, would have been
much the same had Newton never existed. One of the most important but ill-
defined concepts in eighteenth-century science was the 'ether'. The word was
borrowed from Aristotle, but its use in this period seems to have been a legacy of
Cartesian mechanism: the idea of the stars and planets moving freely in empty
space was unacceptable, hence an imperceptible, subtle matter must surround
them. Even Newton accepted the idea of the ether, although he admitted that he
could not describe or conceive of its nature. Huygens used the concept of the
ether to support his novel theory that light travelled not in 'corpuscles', as most
scientists believed, but in waves, which spread out through the ether like ripples
in water. This idea was not widely accepted at the time, but acquired great significance much later, in the nineteenth century. But the ether was generally agreed
to have specific substance, indeed astronomers used its supposed resistance to
account for certain slight discrepancies in celestial motion, that of the Moon for
example.

Despite this Cartesian legacy however, the century and more after Newton's
death provided some remarkable vindications of his theories, the first concerning
the shape of the earth. Newton had predicted that the earth was not a perfect
sphere, but would be found to bulge slightly at the equator under the effects of
centrifugal force. Conversely gravity would be slightly weaker than in northern
latitudes, since the equator was further from the earth's centre. The Cartesian theory predicted the reverse, that the earth would be elongated towards the poles
because the vortices squeezed it at the equator. Rising interest in accurate mapping, especially in France as part of the state programme of science, made it
desirable to determine accurately the figure of the earth. After much scientific
argument, two French expeditions were dispatched in 1735 to measure an arc of
longitude in Lapland and in Peru, under respectively Pierre de Maupertuis and
Charles de la Condamine. The physical diffculties were enormous, especially for
the South American group, whose results were not finally obtained until 1743,
but the conclusion was clear, for in Lapland the longitudinal degree measured
69.04 miles, while in Peru it was 68.32 miles: the earth was therefore considerably flattened towards the poles. On this same expedition, la Condamine's
companion, Pierre Bouger, discovered the magnetic anomalies of different rock
formations which now bear his name, while la Condamine himself undertook a
journey down the Amazon, and it was his later reports which first drew attention
to the zoological and botanical riches of South America.

Among the most effective European advocates of Newtonianism was Voltaire,
who lived in England between 1726 and 1729, where he found social and intel-

SAPIENTI·SSIMI OPUS·

Leonhard Euler's version
of the Cartesian vortices.
Resistance to Newton's
dynamics persisted among
some continental scientists,
and Euler was among
those who adhered to the
view that a physical ether
sustained the heavenly
bodies. Here our solar
system is shown surround-
ed by other such systems
which may be presumed
to exist, each with its own
whirlpool of subtle matter.
From Euler's *Theoria
motuum*, 1744.
The British Library 8561.e.19.

lectual freedoms that were lacking in France; in a famous description he contrasted the dangers of Paris with the conviviality of London by saying that Paris
was full of Cartesian vortices, while London was ruled by Newtonian attraction.
Voltaire's mistress, the Marquise de Chastelet, was a gifted mathematician who
translated the *Principia* into French, and a succession of mathematicians set about
refining Newton's celestial physics. Leonhard Euler (1707–1783) was a native of
Basel who spent much of his life at the Russian court in St Petersburg. Euler
showed that the earth's elliptical path was a varying one, and that the axis of the
earth's orbit around the Sun had shifted by about five degrees since the time of
Ptolemy. The most intractable problem in Newtonian mechanics was the motion
of the Moon as related to that of the Sun and the Earth, the so-called 'three-body
problem', for the Moon appeared subject to slight cyclical variations whose physical cause was baffling, and difficult to reconcile with gravitational theory. Euler
suggested the resistance of the ether as a cause. One of the many outstanding
French mathematicians of this period, Joseph Lagrange (1736–1813), solved
part of the mystery, that of libration: it had long been noticed that, while the

Moon always turns the same face to the earth, there was a small area at each edge of its disk which was visible and invisible by turns. Lagrange showed that the Moon was not a perfect sphere and that the force of gravity therefore did not act truly from its centre, producing this oscillating motion; he predicted that the same would be true of the Earth if seen from the Moon.

Further problems in the Moon's movements were explained by Pierre Laplace (1747–1827), who showed that the Moon would slightly accelerate then decelerate in its orbit, over a long cycle of 24,000 years, and that this was related to changes in the earth's orbit. Newton had recognized perturbations in the planets' paths which he found so puzzling that he supposed that divine intervention was required from time to time to stabilize the system. But Laplace was able to prove that, although the paths of all the bodies in the solar system varied slightly over time, yet they still constituted a highly stable system because all these variations were periodic. As the eccentricity of one orbit increases, so that of another diminishes, in a vast, self-correcting cycle. He showed that there is a plane about which the entire solar system oscillates, and he demonstrated the gravitational effects which each body in the system exerts upon the others. Laplace's immense and forbidding work *Mécanique céleste*, 1799–1825, represents the high-point of Newtonian dynamical astronomy. The famous story is told that Napoleon asked Laplace how such a complex and monumental description of the universe could have avoided once mentioning its author; to which Laplace replied that he had 'no need of that hypothesis'. Laplace was among the first to turn his attention to the origin, in strictly physical terms, of this complex but stable system, and he proposed the nebular hypothesis, that the solar system condensed from a cloud of amorphous gas, a theory which, in a refined form, still prevails today.

Portrait of Pierre Simon Laplace, the great French mathematician who refined and perfected the Newtonian dynamics of the solar system. From the English translation of Laplace's *Mecanique céleste* by Nathaniel Bowditch, 1828.

The British Library 532.i.20-23.

THE PROCESSES OF LIFE

Thus in physics and astronomy Newtonianism triumphed over the mechanical theories of Descartes. But in the life sciences, the image of the organism – and indeed the large-scale processes of nature – as a machine was more tenacious, and had far-reaching consequences, especially in France. The mechanistic model inspired the earliest experiments in the basic physiology of animals and plants. Stephen Hales (1677–1761) was one of that legion of English country clergymen for whom flowers, birds and insects were far more engaging than their parishioners, but Hales was unusual in going beyond description into experiment and quantification. He addressed the fundamental question of plant nutrition, and demonstrated that sap rises with a strong and measurable force, that there is no return comparable to animal circulation, but that the leaves exhaled water vapour. He showed that the leaves 'perform the same office for the support of vegetable life, that the lungs of animals do for the support of animal life', and he recognized that light was essential to plant growth. Hales's book *Vegetable Staticks* was published in 1727, and is the pioneering work of plant physiology. Not until fifty years later did the Dutch scientist Jan Ingenhousz (1730–1799) describe what we now know as photosynthesis: that green plants absorb carbon dioxide, but that this function was restricted to the hours of daylight – that they 'purify the common air in the sunshine, and injure it in the shade and at night'. This discovery was crucial to the idea that plants and animals are linked in a cycle or 'economy' as it was called in the eighteenth century. Contemporary with Ingenhousz was Christian Sprengel (1750–1816) who contributed a further vital discovery about the interrelation of plants and animals, with his demonstration of the role of

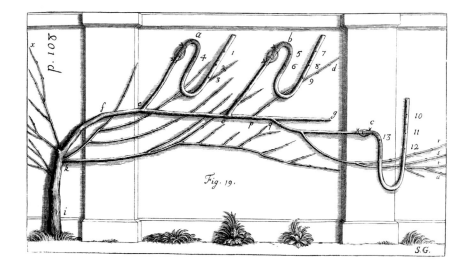

Measuring the pressure of sap in plants. Stephen Hales was the pioneer of plant physiology in the eighteenth century; here he has attached a pressure gauge to the cut tips of a vine. Hales's work *Vegetable Statics*, 1727, was the first to apply such quantitative techniques to plants.
The British Library 1146.d.41.

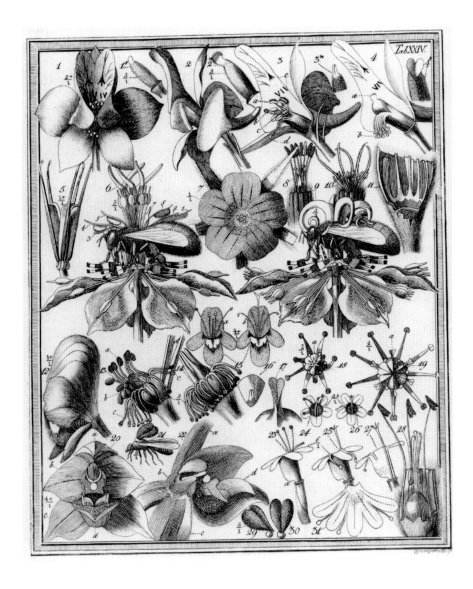

Conrad Christian Sprengel and pollination. Sprengel's detailed observations proved that although most plants are hermaphrodite, they do not self-pollinate, but rely on insects, and that the plant and animal worlds are thus linked in an organic system. From Sprengel's *Geheimnis der Natur*, 1793.
The British Library 449.i.24.

insects in pollinating flowering plants. Sprengel observed very acutely that although flowers are hermaphrodite, the anthers and stigmas mature at different times, so that self-pollination does not occur. Perhaps surprisingly, the old problem of spontaneous generation was revived as late as the 1750s, when John Needham (1713–1781) repeated Redi's experiments with food in sealed containers, but obtained opposite results: he saw tiny organisms swarming and reproducing in the food. It seems certain that the containers which Needham used were not sterile, but the controversy continued until Lazzaro Spallanzani (1729–1799) proved conclusively in 1765 that food hermetically sealed in sterile containers remained free of these tiny life-forms. The idea of spontaneous generation would still be revived from time to time until the era of Pasteur, but scientifically it had been proved a myth. Likewise Charles Bonnet (1720–1793) became convinced that he had proved the reality of preformation and *emboitement,* when he observed, correctly, that female aphids can reproduce without access to males; belief in preformation would also survive into the nineteenth century.

The great representative biologist of the period was Georges Leclerc, Comte de Buffon (1707–1788), an enlightened aristocrat who employed his wealth and his country estate in the pursuit of natural history, and whose vast forty-four volume *Histoire naturelle,* 1749–1804, became a repository of eighteenth-century knowledge and speculation. Its outstanding feature was its conscious abandonment of the doctrine of the fixity of species, and its affirmation that nature was in a state of change. Only thus could Buffon explain the existence of groups of animals which were obviously related to each other – man and ape, horse and ass for example. He went so far as to suggest that these species might share a common ancestor, but rather than progressive evolution, Buffon conceived that degeneration was at work: thus an ape was a degenerate man, the ass a degenerate horse, and so on. A similar process of change was to be seen in the history of the earth, which he suggested had arisen from the collision of a comet with the Sun. Whether as a result of such a collision, or arising from the nebular hypothesis, the cooling of the earth over thousands of years became an accepted principle of geological thought. The cooling of the earth, the formation of rocks and minerals, ocean and atmosphere, and the emergence of life are all described by Buffon as natural processes, quite distinct from any biblical frame of reference, and they occurred, Buffon estimated, over some 100,000 years. Buffon was also a strong supporter of epigenesis, as against the still widespread belief in preformation. Thus Buffon produced a rational, non-Biblical map of nature suited to the age of enlightenment: by rationalizing the relationships between species he prepared, in a general way, for the more mature theories of evolution that were to come.

The most radical denial of the fixity of species came in the work of Jean-Baptiste de Lamarck (1744–1829), who extended the time-scale of the earth far beyond that of his tutor, Buffon. 'Time is insignificant', he wrote, 'and is never a difficulty for nature. It is always at her disposal and represents an unlimited power with which she accomplishes her greatest and smallest tasks.' Lamarck had no doubt that species had evolved and become differentiated over time, and the mechanism which he proposed to explain this was the inheritance of acquired characteristics: for example the long legs of a wading bird or the long neck of a giraffe arose from constant stretching as these animals sought for food in their different environments. Lamarck has often been ridiculed for his supposed over-simplification, but in fact his only mistake lay in imagining that such characteristics could have developed in just two or three generations. Darwin too accepted that variations occured in nature, and it was essential to his theory that these differentiated traits

HISTOIRE NATURELLE.

)OOOOOOOOOOOOOOOOOOOOOOOOOOOOOOOX(

PREMIER DISCOURS.

De la manière d'étudier & de traiter l'Histoire Naturelle.

L'HISTOIRE Naturelle prise dans toute son étendue, est une Histoire immense; elle embrasse tous les objets que nous présente l'Univers. Cette multitude prodigieuse de Quadrupèdes, d'Oiseaux, de Poissons, d'Insectes, de Plantes, de Minéraux, &c. offre à la curiosité de l'esprit humain un vaste spectacle, dont l'ensemble est si grand, qu'il paroît & qu'il est en effet inépuisable

A ij

Opening page of Buffon's *Histoire Naturelle*, 1749. Buffon's vast work advanced the view that the earth and its life-forms were the product of long evolution, and that the biblical account of creation and its time-frame were mythical. In the vignette, the natural and the exact sciences are both invoked.

The British Library 31.d.7.

were inherited, otherwise new species would never become established. These variations or mutations did not occur as simplistically as Lamarck supposed, but the gulf between his ideas and Darwin's has been unjustly exaggerated. Lamarck was an acute observer, and it was he who devised the term 'invertebrate' for the major class of lower animals; he also invented the word 'biology'.

The most permanent achievement of eighteenth-century biology was however not the theorizing of Buffon or Lamarck, but the description and systematic classification of Karl Linnaeus (1707–1778). The number of animal and plant species known to European naturalists was increasingly dramatically as explorers returned with specimens from America, Asia and the Pacific, and it became imperative to bring order into this diversity, to detect patterns and relationships among these innumerable life-forms. Linnaeus was not the first to attempt this, but his system was so clear and comprehensive that it became universally accepted. Linnaeus classified animals and minerals, but his most complete and

CAROLI LINNÆI
Doct. Med. & Acad. Imp. Nat. Cur. Soc.

FLORA
LAPPONICA
Exhibens
PLANTAS
Per
LAPPONIAM
Crescentes, secundum Systema Sexuale
Collectas in Itinere
Impensis
SOC. REG. LITTER. ET SCIENT. SVECIÆ
A. CIƆ IƆ CC XXXII.
Instituto.
Additis
Synonymis, & Locis Natalibus Omnium,
Descriptionibus & Figuris Rariorum,
Viribus Medicatis & Oeconomicis
Plurimarum.

AMSTELÆDAMI,
Apud SALOMONEM SCHOUTEN.
CIƆ IƆ CC XXXVII.

VIRO NOBILISSIMO ET CONSULTISSIMO
D: GEORGIO CLIFFORTIO J. V. D.

Title-page of Linnaeus's *Flora Lapponica*, 1737. As a young man Linnaeus undertook a expedition to Lapland to investigate its flora and fauna; he was equally struck by the Lapp people and the manner in which their lives were regulated by the migration of the reindeer.

The British Library 450.f.l.

influential work was in botany. His plan was announced in the first edition of the *Systema naturae*, 1735, which, in arranging plants according to their sexual characteristics, effected a revolution in the language of botany. The whole plant kingdom was grouped into classes, orders, genera and species. The classes were named after the number of stamens on the plant – *monandria, diandria, triandria* and so on – and each class was then divided into orders, according to the number of pistils, or female parts, in the flower. In identifying each individual plant within these large orders and classes, Linnaeus employed the famous binomial system, the first term for the genus and the second for the species. Thus the goat-willow is *Salix caprea,* the mistletoe *Viscum album,* and so on; in each case the genus is shown by the Latin name, and the species offers some added description – 'goat's', 'white', or, when all else failed, 'common'. This system was so comprehensive and flexible that any new plant could swiftly be slotted into it by picking out the features required by the scheme, and if necessary new orders and sub-orders could be created at will. But of course it was to a large extent an *artificial* system: it was based entirely on a few selected external features, for Linnaeus had no pretensions to advancing plant anatomy or physiology. Above all there is no clear conceptual definition of the crucial words *genus* and *species,* indeed it might be said that the two centuries since Linnaeus have been occupied by the search for such a definition, now taken into the realm of genetics. It is remarkable that a term like *species* should have become so indispensable, yet remain undefined for so long. In zoology, Linnaeus was less successful in his classifications, although he

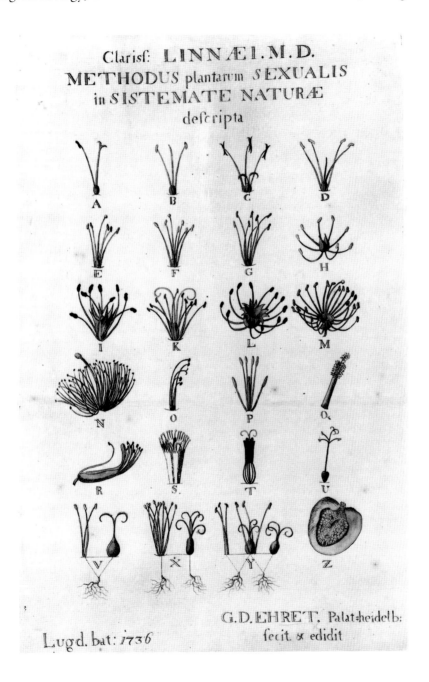

Linnaeus's system of plant classification was announced in his *Systema natura* of 1735. All plants were divided into 24 primary classes according to the number and arrangement of their male organs, the anthers or stamens, monandria having one, diandria two, and so on. All of these classes were then subdivided according to the type of female organs. The Linnaean system was one of the great expressions of the eighteenth-century passion for classification, and it has survived in principle ever since.
The British Library
C.161.d.5.(1)

unhesitatingly placed man among the primates, and in the tenth edition of *Systema naturae* of 1758, he first recognized that whales were mammals. The complexity of animal anatomy made it misleading to select external features as the basis of classification. This tenth edition has become the agreed basis of botanical nomenclature, supplemented and modified since then, but never abandoned.

TABVLA QVARTA.

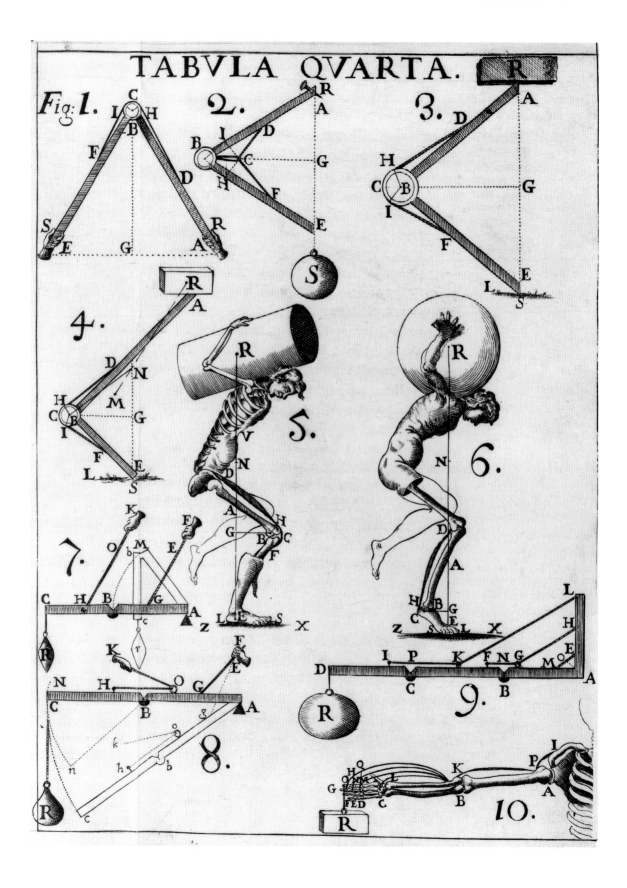

Linnaeus was a religious, conservative man who clung to the doctrine of the fixity of species, and had no time for the speculations of Buffon. He saw the complex balance of nature as clear evidence of a designing God. He described his experience as scientist with these words in the introduction to *Systema naturae*: 'I saw the infinite, all-knowing and all-powerful God from behind as he went away and I grew dizzy. I followed his footsteps over nature's fields, and saw everywhere an eternal wisdom and power, an inscrutable perfection.' This was the age of the great botanical folios with their superb illustrations, such as those by Georg Ehret, where formal precision was allied to an artistic beauty which seemed to invite the reader to enter a scientific Eden.

The dilemma of the life-sciences in the eighteenth century was reflected in the medicine of the period. There was an impressive accumulation of data, but it was coupled with a deep uncertainty about basic physiological processes and about the causes of disease. The anatomical map of the human body became more and more detailed through the work of teachers such as Albrecht von Haller (1708–1777) of Bern, whose *Icones anatomicae* was probably the most meticulous work of its kind since Vesalius. Yet despite his extensive experiments in embryology, Haller adhered to the theory of preformation, denying the possibility of epigenesis. Haller emphasized the different kinds of tissue that form the body, and made a special study of muscular action, which reacts in ways quite distinct from other tissue. He coined the term 'irritability' for its power of contraction, but he did not believe that this contraction was controlled by the nerves. Among the greatest theoretical divides in eighteenth-century medicine was that between the mechanists and the animists. Haller did not accept the Cartesian model of the body as a mechanism, which was so attractive to many physicians, and which was propagated by the leading pedagogue of the age, Hermann Boerhaave of Leiden (1668–1738). This model was given its most radical expression by one of Boerhaave's pupils, Julien La Mettrie (1709–1751), whose book *L'Homme Machine* of 1748 was a sensational statement of materialism and atheism. La Mettrie argues that all psychological states as well as bodily functions are caused by physical stimuli and by the brain structure. His other notorious work *L'Histoire naturelle de l'âme*, 1745, likewise refuted the traditional doctrine of the spiritual and immortal soul. By no means all mechanists accepted La Mettrie's extreme statement of the Cartesian position, but since the chemical approach to physiology lay far in the future, what alternatives were there to this mechanical model? The leading opposition came from the school of 'vitalism' or 'animism', which acknowledged that the life-functions were beyond physical analysis, and were directed by some vital principle, often termed the *anima,* and therefore analogous in some way to soul or spirit. Haller was drawn to this position, and speculated whether the *anima* resided in the blood, as traditionally believed, or in the nervous system. He was deeply impressed by a study he made of Siamese twins, who shared a heart and a blood system but evidently had two distinct *animae,* two life-spirits, two wills.

Vitalism as a doctrine is associated most often with Georg Stahl (1660–1734) who argued that the mechanisms of the body were merely instruments directed by the *anima*. He held that this could be proved by the physical effects caused by psychic states – fear, desire, anger and so on – effects which were inexplicable in purely mechanistic terms. Stahl was equally famous as the leading proponent of the theory of 'phlogiston', which explained combustion by supposing that this elementary combustible substance was to be found in virtually all matter, and that it was consumed during burning. This theory had enormous commonsense appeal, and the awkward fact that the calx left after burning is almost always heavier than the original material was explained by arguing that phlogiston had

In his work of 1680, the Italian physicist Giovanni Borelli, analysed the movements of the human body in terms of precise mechanics, with the arm and leg joints acting as levers and pulleys. Such analyses reinforced the Cartesian view of man as a machine.
The British Library 38.c.12.

Statue of Edward Jenner vaccinating a child against smallpox. Jenner's conquest of smallpox, announced in 1798, was the single greatest advance of its era in practical medicine, and it made him an international hero: this commemorative statue was made in Italy in 1873.

Galleria Nazionale del Arte Moderna, Rome

Muscle system from Bernhard Albinus's *Tables of the Skeleton and Muscles of the Human Body,* the English version of a massive atlas of anatomy first puiblished in Latin in 1747. Albinus worked tirelessly with the Dutch artist Jan Wandelaar to achieve these meticulous pictures with their idealized classical backgrounds.

The British Library 3 TAB 73 (1)

'negative weight', that is, some form of buoyancy, whose disappearance left the material heavier.

The tension between mechanists and vitalists illustrates both the strength and the limitation of eighteenth-century science: answers about the fundamental nature of life were felt to be within the grasp of rational thought, yet crucial gaps in scientific knowledge (above all perhaps in chemistry) compelled biologists to appeal to entities that were theoretical and unverifiable. It is remarkable that in an age which we regard as above all rationalist and empiricist, a controversy such as that concerning preformation and epigenesis should be incapable of resolution. The really decisive observations that refuted preformation were published by Caspar Wolff (1734–1794) in the 1760s and 1770s. Wolff was at last able to chart convincingly the appearance and growth in embryos of the heart and blood vessels from previously undifferentiated tissue. The idea of life as a specific, mysterious yet tangible force reached its clearest expression in Luigi Galvani's doctrine of 'animal electricity'. Galvani (1737–1798) believed that his experiments (see below, page 178) proved that living tissue contained a subtle fluid related to electricity, and in some quarters the link was made between electricity and life, an idea which lingered long in the imagination, as *Frankenstein* and other scientific fictions demonstrated.

The most impressive single advance in eighteenth century medicine was without doubt the technique of smallpox vaccination, pioneered by Edward Jenner (1749–1823). Smallpox had replaced bubonic plague as the greatest health scourge in Europe, and when it failed to kill, it left most disfiguring scars. Jenner was a country physician with no pretensions to the philosophical high-ground occupied by figures like Stahl or Haller. His discovery was the result of simple observation and deduction: that farm-workers, and especially milkmaids, who had contracted the mild complaint known as cow-pox, were apparently immune to smallpox. Inoculation with smallpox itself had been tried, but was often found to be fatal. By taking a tiny sample from a cow-pox sore and scratching it into the skin, Jenner produced immunity to smallpox itself at the cost of a mild attack of cow-pox. His publication of 1798 announcing this fact electrified the international medical world, and his results were soon strikingly proved. National vaccination programmes were set up, Jenner became a hero, and in 1804 Napoleon had a medal struck in his honour, notwithstanding the fact the Britain and France were at war. It cannot be said however that either Jenner or his contemporaries went on to formulate new theories of disease, and Jenner himself returned to his studies of bird-life, becoming the first to identify the strange habit of the cuckoo, which would destroy the eggs of its host by forcing them out of the nest. Jenner's achievement did however enormously boost the prestige of medical science, which, as eighteenth-century art and literature can testify, was still very much a collection of hit-and-miss techniques. By the end of the century the careers of men such as the London surgeon John Hunter (1728–1793) had raised surgery to a much higher intellectual and social level than it had enjoyed before. By performing post-mortems, Hunter was able to evaluate the progress of diseases and the effect of treatments. His great collection of biological specimens became a national museum, and he was one of the founders of systematic pathology.

Between empirical study and fanciful speculation, new models of the body's functions were proposed and evaluated, and the same was true of the earth itself: as the century progressed, the idea that the earth had been formed by slow, powerful natural processes gained wide acceptance, because they could be seen to be

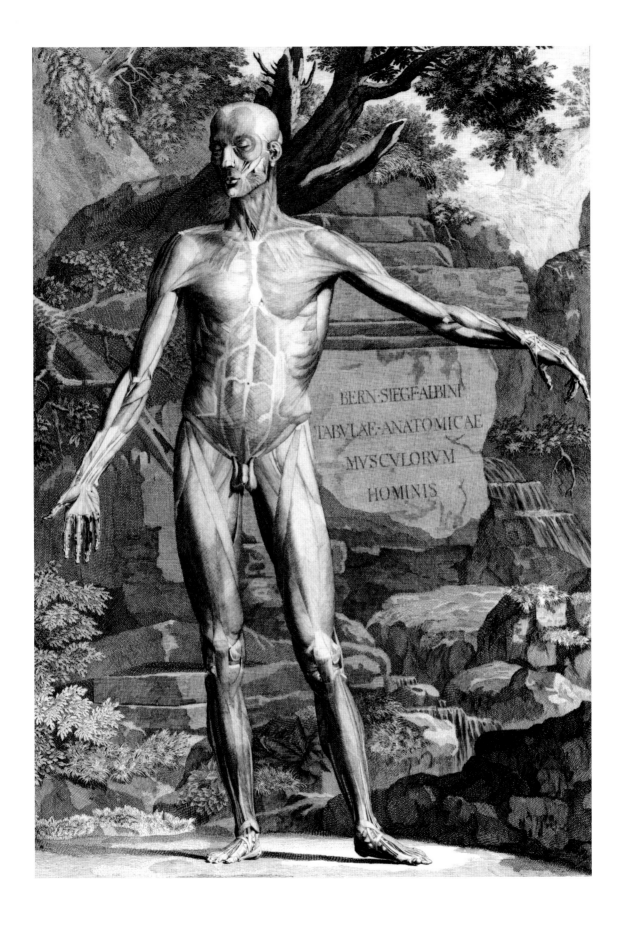

BERN·SIEGF·ALBINI
TABVLAE·ANATOMICAE
MVSCVLORVM
HOMINIS

still at work. At first these processes were conceived to have occurred within a biblical time-frame of a few thousand years, but writers such as Buffon were soon compelled to extend this as they considered processes like mountain-building, sedimentation of river estuaries or the erosion of coasts. One of the most comprehensive early collections of fossils was made by the Swiss, Johann Scheuchzer (1672–1733), built up during the numerous excursions he made into the Alps. Scheuchzer was convinced that fossils were of organic origin, an idea which was still controversial, despite the pioneering statements of Stensen, but he took their presence high in the mountains to be proof of the biblical flood. Yet his fieldwork and the richness of his collection makes Scheuchzer one of the founders of palaeontology.

The Swiss naturalist Johann Scheuchzer collecting fossils in the Alps. Whether fossils were purely geological objects or had animal origins was much debated. Scheuchzer argued strongly that they were the remains of vanished life-forms, many of them destroyed in the biblical flood, thus explaining their presence in high mountains far from the sea. From Scheuchzer's *Museum diluvianum*, 1716.

The British Library C.262.(1)

One of the major theories of early geology became known as Neptunism: the belief that all rocks, including the ancient crystalline forms such as granite, had been precipitated from water, and that therefore ocean had once covered all the earth and had gradually retreated. The most famous proponent of this theory was Abraham Werner (1749–1817), a very influential teacher at a mining academy in Freiberg, Saxony. In Werner's thought, the stratified rocks containing fossils were laid down later as the ocean receded, and the volcanic rocks were the latest of all. A contrary view was taken by the Plutonist school, sometimes known also as the Vulcanist, who saw volcanic disturbance as the primary force that had formed the earth's crust. In this theory subterranean heat threw up landmasses, which were in turn worn down by erosion, and this process was repeated in a never-ending cycle. The leader of the early Vulcanists was James Hutton (1726–1797) whose *Theory of the Earth,* 1795, was the most influential geological work before Lyell. Hutton recognized that rocks such as granite were igneous in origin, and his views were more modern and closer to the truth than those of the Neptunists – Werner could not, for example, explain what had become of the oceans that had receded across the world.

A related problem was the hydrological cycle: from where did rivers arise, and why did river- and rain-water not cause the sea to rise and flood the land? Many eighteenth-century theorists were attracted to the idea of a subterranean world, through which sea-water flowed in hidden channels back to the sources of rivers (losing its salinity in the process). But Edmond Halley was able to show that rivers, seas and rain were linked in a cycle, by calculating the discharge of Europe's principal rivers into the Mediterranean, and linking this to the water lost by evaporation each day. He deduced correctly that the evaporated water condensed and fell as rain, which swelled the rivers and returned again to the sea. This quantitative approach was capable of yielding very significant results of this kind, but in the eighteenth century it was always likely to be supplemented by rather wild speculation. Attempts would be made well into the nineteenth century to harmonize the newer geological thought with scripture. Thus Georges Cuvier (1769–1832), the most significant palaeontologist of his age, considered that the gaps in the fossil record were caused by catastrophic events in the earth's history, which had periodically destroyed all life, which must therefore have been restored by 'successive creations'. This theory was gratefully seized upon by geologists wishing to re-affirm the traditional religious framework, such as William Buckland in Oxford in the 1820s, who identified the latest of the catastrophic events with the biblical flood, and thus made geological science acceptable, for a time at least, to conservative religious thinkers.

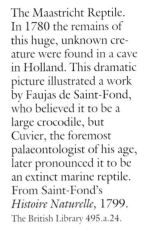

The Maastricht Reptile. In 1780 the remains of this huge, unknown creature were found in a cave in Holland. This dramatic picture illustrated a work by Faujas de Saint-Fond, who believed it to be a large crocodile, but Cuvier, the foremost palaeontologist of his age, later pronounced it to be an extinct marine reptile. From Saint-Fond's *Histoire Naturelle*, 1799. The British Library 495.a.24.

ELECTRICITY AND INSTRUMENTATION

The difficulty for scientists working on the history of the earth and its life-forms was that a traditional religious account of their origins had permeated western culture, an account which must be either harmonized with contemporary knowledge or painfully rejected. No such framework troubled the most original scientific discovery of the eighteenth century, namely electricity. Several early experimenters, William Gilbert and Otto von Guericke for example, had discovered that a metallic sphere, when rotated and rubbed, acquired the power to attract small objects and to give off sparks. But it was in the 1720s that interest in electricity really developed, when two English experimenters, Francis Hauksbee (1666–1713) and Stephen Gray (1666–1736), built more effective static generators – globes of evacuated glass turned on a spindle, which produced sparks or a glowing light, and made certain objects attract each other. Gray made the significant discovery that an electric force could be carried for hundreds of feet along some materials, such as silk, but which refused to be transmitted through certain other substances, or if the medium were too thick, such as heavy brass wire: thus both conductivity and resistance were discovered. Amusing experiments were devised in which an electric current was carried in this way across a room and used to ignite alcoholic vapours. Hauksbee and Gray could only use traditional language to describe the force they were investigating – 'virtue' or 'subtle fluid' – with the implication that electricity had a real, physical existence within material objects, which could be released under the right conditions. There was considerable excitement about the nature of this strange force. Just before his death, Gray had discovered that a conical pendulum would revolve about an electrified body, and he made an instant connection between this movement and the motion of planets about the Sun. He hoped 'if God would spare his life but a little longer..to astonish the world with a new sort of Planetarium … and a certain theory for accounting for the motions of the Grand Planetarium of the Universe'.

A decisive step towards isolating this force occurred in 1746 when Petrus van Musschenbroek of Leiden (1692–1761) determined to test whether the electric 'fluid' might be stored like any other. A static generator was connected to a bottle of water coated with metal, which thus became charged; when Musschenbroek touched the metal leading into the bottle he received so strong a shock that he vowed he would not repeat the experience 'for all the kingdom of France'. The

The Leyden Jar, an electrical condenser, discovered by van Musschenbroek in 1746. Static electricity is being generated by the revolving glass sphere on the right, and conducted into the jar of water on the left. When Musschenbroek withdrew the contact from the water, he received a severe shock. The background to the picture evokes the wonder of the newly-discovered force, and suggests that electricity consists of a stream of subtle matter. From Nollet's *Essai sur l'électricité des corps*, 1746.
The British Library 1489.h.36.

report of the 'Leiden jar' (which would later be termed a condenser or capacitor) spread rapidly, and its stored power facilitated further experiments, as well as salon games in which circles of men and women, hand in hand, were electrified. In America, Benjamin Franklin (1706–1790) discovered that lightning was natural electricity by drawing down a lightning charge into a Leiden jar. Franklin's result showed that electricity was a natural force of immense power, with implications that went far beyond mere parlour games. One scientist who eagerly repeated Franklin's experiment with lightning was Georg Richman (1711–1753), who had been trying for some years to devise a means of measuring electric forces. Richman set up lightning conductors in his home in St

The death of Georg Richman. Early experimenters with electricity scarcely realized the dangers of the new force. During a thunderstorm in St.Petersburg in 1753, Richman was repeating Franklin's exploit of drawing down lightning, but he had neglected to insulate himself from the earth, and he was killed by a violent discharge. This dramatic picture is not contemporary, but is a nineteenth-century artist's view of the event.

Petersburg, and during a summer thunderstorm in 1753, he was killed when reading his 'electrometer', an event which shocked the scientific world. The first successful attempt to measure electricity came when the French engineer Charles Coulomb (1736–1806) devised a 'torsion balance' which showed the force required to bring two electrically-charged, and therefore mutually repellent, objects together. Coulomb's results were of enormous interest because they showed that the same inverse-square law which Newton had discovered in the case of gravity was true of electrical attraction: the force varied as the square of the distance between the charged and uncharged bodies.

All these experiments down to the 1790s had made use of static electricity, which was the only form then known, while the discovery of current electricity came about by chance. Luigi Galvani's theory of 'animal electricity' arose from his discovery that the muscles of dead frog would twitch when the nerves were touched with an electric probe. Galvani's contemporary, Allesandro Volta (1745–1827), repeated Galvani's experiments and reached a very different conclusion. Volta found that the muscular contractions were actually the result of the

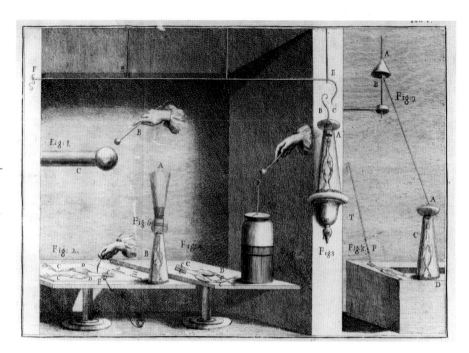

Galvanism or animal electricity: Luigi Galvani found that the legs of frog would twitch if contacted with an electrostatic generator, or with two different metals in a damp medium. He concluded that 'animal electricity' was intimately connected with life itself, an idea which persisted for long after Volta gave another explanation of these effects. From Galvani's *De viribus electricitatis*, 1791.
The British Library 435.f.22.

contact across a damp medium between the different metals of the probe and the plates on which the frogs were mounted. He verified that further small electrical effects would result from two metals in a damp environment, for example, a tingling on the tongue from a bi-metallic strip. Conclusive proof was obtained when he constructed his 'Voltaic Pile': a series of discs, alternately zinc and copper, were separated by paper soaked in brine or weak acid, but connected by a conducting strip. This pile produced a steady electric current, which could be strengthened by joining further such piles in series. This was the first cell or battery, described by Volta in 1800, and the only possible conclusion was that some chemical reaction between the metals was responsible for the current. Great interest was aroused when such a pile was partly immersed in water, and bubbles of gas were seen rising from the metals, which proved when collected to be hydrogen and oxygen: the water gradually disappeared, for it was being decomposed into its constituent gases. The implications for the theory of matter were deeply puzzling, and would necessitate a new synthesis of physics and chemistry, although this was many years in the future.

Volta is one of the several eighteenth-century scientists whose names were later commemorated as measurement and instrumentation became more important and more precise: Volta, Ampère and Coulomb in electricity, Fahrenheit and

Celsius in heat. Forces such as heat, light and sound presented the most difficult challenge to physicists, for they were believed to consist of particles, yet they resisted measurement, and hence had formerly been described as 'imponderable' fluids, that is, incapable of being weighed. Thermometers had been made in the seventeenth century, but their scales were arbitrary and it took some time to develop the principle that agreed upper and lower fixed points were essential if the instrument was to be scientifically used. Daniel Fahrenheit (1686–1736) was a Prussian instrument-maker who in 1708 visited the Danish astronomer Ole Rømer and watched him making thermometers from spirit confined in a narrow tube, using ice and blood-heat as the two extremes. Rømer died without publishing his methods, but Fahrenheit continued his work, using mercury as the indicator, taking 90 as the blood temperature and 30 as the freezing temperature, which suggests perhaps that he had in mind meteorological use in Northern Europe. He soon discovered that water boiled at 212 on this scale, and his system was complete. An alternative scale was devised by the Swede, Anders Celsius (1701–1744), but curiously he took 0 as the boiling point of water, and 100 as its freezing point. It was Linnaeus who suggested that it was more logical to reverse the scale, and in this form it became widely used.

   Instrument-makers such as Fahrenheit were rarely the philosophers of science, but they often worked at the frontiers of experimentation, and the measurement of temperature had far-reaching consequences, for the Scots physician Joseph Black (1728–1799) could scarcely have pioneered the science of heat without these agreed scales. Black was able to show that the heat required to produce a given rise in temperature was not the same for all bodies, and that materials therefore had different 'specific heats'. Temperature was therefore quite distinct from heat. Black also devised the concept of 'latent heat', where heat may change the state of matter, for example from liquid to gas, without changing its temperature; the reversibility of this process would have vital consequences for steam engineering. Other instrument-makers whose practical genius contributed enormously to the exact sciences were John Harrison (1692–1776), whose chronometer revolutionized navigation, allowing longitude to be determined, and Jesse Ramsden (1732–1800) whose equatorial mounting coupled a clockwork mechanism with a telescope, enabling any point in the heavens to be held in view.

   The drive towards measurement in the eighteenth century is perhaps best seen in the inauguration of the decimal system. The reform of the innumerable customary units of weights and measures had been discussed in Europe for a century or more before the French Revolution provided the psychological impetus to replace the old systems. In 1790 the Academy of Science recommended to the National Assembly that the earth itself should be used as the basis for a rational measurement system. The meridian from the North Pole to the Equator should be measured, and one ten-millionth part of that distance should be termed a metre, with fractions and multiples of it having rational prefixes derived from Greek. King Louis XVI issued the proclamation to begin the work from his prison cell, and the two chief architects were to be Jean-Baptiste Delambre (1742–1822) and Pierre Méchain (1744–1804). In the fog of war and revolution, the two scientists laboured for six years to measure an arc from Barcelona to Dunkirk, before the metre could be defined. The new system was finally enacted in France in 1799, 'for all people for all time'. Simultaneously the new standard of weight, the gram, was made equal to the weight of a cubic centimetre of pure water, a definitive kilogram was manufactured from platinum, and a litre was defined as the volume of a cube whose sides were all 10 centimetres. French political dominance during these years carried the new system throughout Europe, but inevitably its universal acceptance required two or three generations – no doubt in accordance with Lamarckian evolution.

NATURE, GOD AND MAN

In 1714 William Derham wrote of Newtonian astronomy: 'This system is the most rational and probable because it is far the most magnificent of any, and worthy of an infinite Creator.' Derham (1657–1735) was an Anglican clergyman, socially well-connected, a competent amateur scientist and a fellow of the Royal Society. His two original works, *Physico-Theology* (1713) and *Astro-Theology* (1714) took as their theme the magnificence of God as shown in the glories of the natural world and the cosmic structure. The possibility that the universe was infinite, which had been so radical and dangerous a belief in the time of Copernicus, Bruno and Galileo, is eagerly embraced by Derham because it would the more magnificently display God's creative power. He accepts that every fixed star is a sun, probably with attendant planets, which in turn probably support life as ours does. The novelty of this belief is worth dwelling on: for centuries Christian thought had taken it that the universe existed merely as a theatre for man's fall and redemption. The idea of the 'plurality of worlds', as it was called in Derham's day, could surely be seen as tearing down the walls of this theatre, as undercutting the unique personal relationship between God and world of man that had always been at the heart of religious faith. Yet Derham had no uneasiness about the implications of his ideas, for the new vistas opened by science were smoothly transformed into a supremely rational religion: beneath all the diversity of nature lay laws so powerful that any new demonstration of their complexity could be welcomed as further proof of God's omnipotence. This synthesis of religion and science was to prevail in English thought throughout the eighteenth century and well into the nineteenth. The design that was everywhere apparent in the universe was seen as proof not merely of God's existence, but of his wisdom and goodness.

Thirty years after Derham's death, Baron d'Holbach wrote, in a France where social and intellectual tensions were becoming intolerable, 'Theology is but the ignorance of natural causes reduced to a system', and again 'When man ascribes to 'the gods' the production of some phenomenon … does he in fact do anything more than substitute for the darkness of his own mind, a sound to which he has been accustomed to listen with reverential awe?'. Among the group centred around Diderot and the *Encyclopédie,* d'Holbach (1723–1789) was perhaps the most strident atheist. To him, nature was a self-sustaining mechanism which needed no controlling intelligence, and science as the study of nature was a purely secular pursuit. God was merely a label for man's ignorance of nature's secrets, and as science progressed, it would inevitably demythologize the universe. Behind this confidence lay the Cartesian model of the mechanism, whose workings would eventually yield to rational investigation. The French philosophers were especially concerned by the implications of this model for the understanding of man. If man was a natural entity, to be analysed and interrelated with other forms of life, then science acquired a political dimension, for whence arose the distinctions of rank and power with which human society was riddled? The discovery of the 'noble savage' in the Pacific suggested the possibility that man had originally existed in a state of freedom and equality. If so, then the historical progress towards civilized society could be seen as a conspiracy to concentrate power in hands of an elite, a conspiracy aided by religious ideas of duty and obedience. Thus, in the hands of these French thinkers, the scientific view of nature was believed to reveal powerful, long-hidden and politically-charged truths about man himself. Science thus became conscious of its own power to criticise and destroy traditional beliefs. It acquired an intellectual authority in its own right, an authority that was secular, radical and exclusive of other approaches to truth. This model arose first amid the social and political tensions of eighteenth-century France; it found little favour in England or Germany, but despite the efforts of deists and rational theologians, it was the model of science that would ultimately prevail.

Some aspects of eighteenth-century science were rigorous, order-producing and of permanent value, such as the mathematics of Laplace, the taxonomy of Linnaeus, or the embryology of Wolff. Yet it was also productive of fanciful theories – ether, phlogiston, animal electricity, and so on – which had no basis in fact, but which commanded widespread assent because they appeared to explain certain phenomena. In Germany a school of thought arose which centred on the supposed relationship between human and divine reason. 'Nature Philosophy' commanded the support of some of the greatest intellects – Kant, Schelling and Goethe – who all emphasized the organic wholeness of nature, and in this it was clearly related to Romanticism. Its influence was felt long into the nineteenth century; in biology it reinforced the vitalist tradition, while in physics it prompted scientists such as Oersted to search for the link between electricity and magnetism. The hallmark of its language was the word *Ur* – meaning original, primeval or archetypal: its philosophers spoke of ur-substances, ur-life-forms, ur-rocks and so on, from which the diversity of nature had evolved. The eighteenth century also witnessed the birth of pseudo-sciences like Mesmerism and physiognomy, to replace the now-outmoded ones, astrology and alchemy. Hypothetical entities such as animal electricity flourished because, outside mathematics and astronomy, there was as yet no true consensus as to the fundamentals of science. Empiricism and experimentation were accepted to be the true basis of science, but the data they furnished must be interpreted, and placed within a conceptual framework. One of the most interesting figures in this context was Emanuel Swedenborg (1688–1772), who researched for many years in geology and palaeontology, and was a convinced Cartesian in his view of the earth and the heavens. But Swedenborg became increasingly concerned with the body-soul problem, and sought to establish a physiology that would embrace the workings of mind, will and spirit. Finding no adequate scientific guidance, Swedenborg veered into the language of the visionary and the mystic, in which however the relationship between the physical and the spiritual world were always of paramount interest. Swedenborg's career sprang in a sense from the immensely ambitious programme of Cartesian science, which was to provide an all-embracing analysis of nature's mechanisms; when that promise proved incapable of fulfilment, Swedenborg retreated into the realm of religious imagination. The limitations of eighteenth-century science, and the striking differences between the national schools of scientific philosophy, remind us that science is not pure knowledge but is a provisional report, a system of beliefs which are historically conditioned. The language of science does not ultimately explain, but re-states accounts of nature in terms convincing to its age.

```
┌─────────────────────────────────────────────┐
│ ┌─────────────────────────────────────────┐ │
│ │                                         │ │
│ │            Chapter Seven                │ │
│ │                                         │ │
│ │          THE MACHINE AGE                │ │
│ │                                         │ │
│ │                 ❧                       │ │
│ │                                         │ │
│ │   'We claim, and we shall wrest from    │ │
│ │      theology, the entire domain of     │ │
│ │         cosmological theory.'           │ │
│ │                          John Tyndall   │ │
│ │                                         │ │
│ └─────────────────────────────────────────┘ │
└─────────────────────────────────────────────┘
```

Chapter Seven

THE MACHINE AGE

❧

'We claim, and we shall wrest from theology,
the entire domain of cosmological theory.'

John Tyndall

THE INDUSTRIAL REVOLUTION

In Shakespeare's England, Francis Bacon had proclaimed that knowledge was power, and had foreseen a society transformed by scientific knowledge, by man's control of natural forces. In its own day this prophecy had rested on very slender foundations, and its fulfilment was long delayed: Galileo, Harvey, Descartes, Newton, Buffon and Laplace had opened intellectual horizons undreamed of by Bacon, but the material basis of society remained unchanged. It was the Industrial Revolution in the years 1780 to 1830 which transformed science into a social force through the agency of technology. The machine affected millions of lives, it made science visible, and it proved the truth of the scientific principle that the world operated by laws and mechanisms, and that these might be discovered and harnessed by man. When Macaulay penned his famous eulogy on science in 1860, it was really technology that he had in mind, for it was the machine which had 'furnished new arms to the warrior ... spanned great rivers and estuaries ... lightened up the night with the splendour of the day ... multiplied the power of human muscles ... accelerated motion, annihilated distance ... and enabled man to descend to the depths of the sea and to soar into the air.'

It is one of the strangest paradoxes in the history of science that the machine culture dawned and gathered pace in apparent isolation from intellectual science. The quintessential early machine was the steam engine – or more accurately the atmospheric engine – but its first builders had no known background in physics or scientific theory of any kind. On the other hand it is inconceivable that Thomas Savery (1650–1715), builder of the first steam pump, was not aware of the principle of the vacuum, or that one could be created by condensing steam. Savery's pump was designed to drain water from mines and was patented as early as 1698. It had no moving parts other than valves, which had to be operated by hand. Steam from a boiler filled a pipe which was then cooled externally by cold water. The resulting partial vacuum would draw up water from the mine, but only to a maximum height of around ten metres because of atmospheric pressure. It was claimed that the cycle of operation could be repeated five times per minute. As far as we know, only four engines of this type were built, and there is no record of how effectively they worked, so that in itself the Savery engine would not have played a great part in technological history had it not inspired other, more sophisticated machines. That designed by Thomas Newcomen (1663–1729) worked on the same principle as Savery's, but the vacuum was made to work a piston,

Plate XXXVII.

front. p. 490.

Newcomen's beam engine, demonstrated as early as 1712, and built by the hundred in eighteenth-century England.
Its working stroke was produced by the partial vacuum arising from condensed steam, thus it was really an
'atmospheric engine' rather than a steam engine, but it marked the beginning of a power revolution
that transformed industry and society. From Desaguliers's *Course of Experimental Philosophy*, 1744.

The British Library 1651/1101

connected to a large horizontal beam, which in turn raised and lowered the pump rod. The hand-operated valves were eliminated, and the work-cycle was repeated up to twelve times per minute. First demonstrated in 1712, many hundreds of Newcomen beam engines were built in England during the next eighty years, yet Newcomen published no account of their working, and he had no known contact with the world of science.

In modern engineering terms, the Newcomen engine was desperately inefficient, because it was alternately heated and cooled, wasting most of its huge input of fuel – critics observed that it took the product of an iron mine to build one engine, and the product of a coal mine to fuel it. This was the central problem which was brilliantly solved in 1765 by James Watt (1736–1819), who added a second chamber to condense the steam, so that the piston chamber could be kept hot. This saved three-quarters of the fuel used by a Newcomen engine. Watt also introduced the double-action principle by using steam pressure to force the piston up, and he was able to transform the reciprocating action of the beam to rotary action by gearing the beam to a huge flywheel. Watt's innovations resulted in a vastly improved power-source, applicable to many more uses than pumping mines. They could be used to turn textile machines, mills and boring machines, to blow air into furnaces and to operate metalworking hammers, rollers and lathes. Steam engines were so prevalent by the turn of the century that it became necessary to grade their power output, and the unit of horsepower was defined as the ability to raise 33,000 pounds weight one foot in one minute against the force of gravity.

Other revolutions in technology were equally vital in the eighteenth century, but perhaps less conspicuous and exciting than these new machines. The smelting of high-quality iron using coal rather than charcoal was pioneered by the Darby family, while the spinning machines of Arkwright and Hargreaves revolutionized the textile industry. One product of new technology that was eminently visible in polite society was the pottery and porcelain that bore the names Wedgwood, Meissen or Sèvres, the result of long experiments with clays and glazes. In all these fields however, scientific theory played virtually no part: it was experiment and craft ingenuity that were at work, driven by the search for commercial gain and for improvement for its own sake. But the crucial point is that once these machines and processes were in operation, they in turn stimulated scientists to study and understand what principles and physical forces were involved, and how machines could be measured for efficiency, and so perhaps improved. This could best be done by developing the theoretical language of physics and chemistry, and thus technology began to interract with science. Both acquired a renewed momentum, driven by intellectual curiosity and by a growing consciousness of their commercial potential.

THE CHEMICAL REVOLUTION

Perhaps the greatest intellectual challenge lay with chemistry – to free it from its alchemical roots and create a logical framework which would bring order to the myriad forms of matter. Chemists had been accumulating isolated facts and insights for centuries, but they remained unrelated, unexplained by any theory of the processes involved, unless they were magical ones. The difficulties were enormous, measuring and weighing substances that were often invisible and intangible, and painfully evolving a language of fundamental concepts, such as element, compound and reaction. Combustion was still seen the main agent of change, and a number of pioneers tried to rationalize its effects, step by step. At Glasgow University, Joseph Black, the pioneer of heat theory, heated chalk and magnesium alba, and collected the resulting gas – carbon dioxide – which he named 'fixed air', because it was evidently locked inside these substances. He was intrigued to find that the process was reversible if the calx of chalk or magnesium was exposed to the air, suggesting that normal air comprised a mixture of different gases. Black used precise balances to show loss and gain in combustion, and he found the same 'fixed air' to be produced during fermentation and respiration. The wealthy recluse Henry Cavendish (1731–1810) – 'the richest of all learned men, and the most learned of the rich' – anticipated Coulomb's inverse square discovery in electricity, but did not publish his findings. He isolated an 'inflammable air' – hydrogen – which was given off by the action of acids on certain metals: was this perhaps phlogiston? Apparently not, since the substances which could release it were not themselves readily combustible. Yet when pure hydrogen was burned together with pure oxygen, a water residue remained, driving Cavendish to the view that water was somehow composed of these two gases. Joseph Priestley (1733–1804) was unusual in England in uniting science with political radicalism, so much so that he was driven to emigrate to America because of his sympathy with the French Revolution. Priestley may be called the discoverer of soda water, after passing 'fixed air' (carbon dioxide) through water, and finding the result very agreeable to drink. But more important, he isolated oxygen by heating mercuric oxide and collecting the gas. He called it 'dephlogisticated air' because it burned brightly. Priestley never abandoned his belief in phlogiston, which handicapped his efforts to rationalize his results. He established that plants made stale air respirable once more by exhaling this same gas.

By their use of the balance and of careful measurement, these chemists of the late eighteenth century established the important principle that matter is conserved during chemical reactions, despite the apparently destructive power of fire or acid; the forms of matter were changed, but the weight of its component parts remained constant. Priestley travelled to Paris in 1774 where he met the great Antoine Lavoisier (1743–1794) and told him about dephlogisticated air. Lavoisier too espoused the doctrine of the conservation of matter, proved by the simple experiment of condensing steam and finding it equal to the weight of evaporated water. He repeated Priestley's experiment to obtain oxygen, and then burned various substances in its presence and in air. Meticulous measurements led him to the revolutionary conclusion that, during combustion, substances combine with oxygen, and that the phlogiston theory was an illusion. Lavoisier stated his belief that all matter was composed of a small number of pure, elementary substances which could not be further subdivided, and that these elements combined into an almost limitless variety of compounds. 'In all the operations of art and nature', he wrote, 'nothing is created: the quality and quantity of the elements remain the same, and nothing takes place beyond changes in the combination of these elements.' In his classic work *Traité élémentaire de chimie*, 1789, Lavoisier enumerated twenty-three of the modern elements, although he also added certain others such as 'caloric', the supposed heat-fluid, and he regarded light as an element. Lavoisier thus finally replaced the classic four-element concept with a table of experimentally-derived elements. Equally fundamental was his reform of chemical nomenclature: he proposed that the names of chemical substances should embody their origin or function. Thus oxygen means acid-creator, based on Lavoisier's belief that oxygen was a constituent of all acids, while ancient alchemical names such as aqua fortis and oil of vitriol became respectively nitric and sulphuric acid. Recognizing that water can be decomposed into oxygen and another constituent, he named that constituent hydrogen – 'water-creator'. This new language was logical and self-consistent, and its creation places Lavoisier as the founder of modern chemistry. He applied

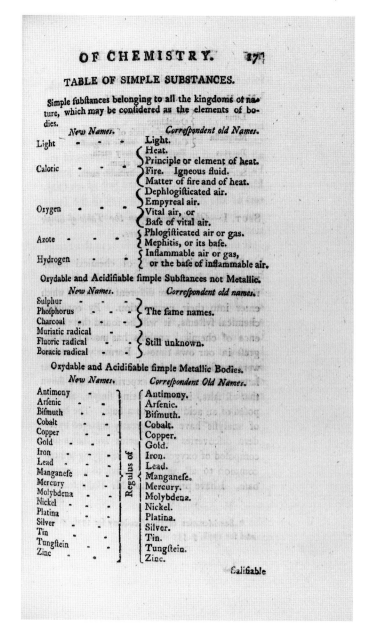

Right: Lavoisier's table of elements from his *Elements of Chemistry*, 1790. The name oxygen was coined by Lavoisier because he believed that it was a constituent of all acids – *oxys* in Greek. *Azote* – 'lifeless air' – is nitrogen. These gases were formerly considered to be forms of the air, but Lavoisier knew them to be elements. However he also believed light and heat to be elements too.
The British Library 1035.g.16.

Facing page: Dalton's table of elements from his *New System of Chemical Philosophy*, 1808. Dalton's unique contribution was to determine the atomic weights of twenty elements, using that of the lightest, hydrogen, as base 1. He devised a set of graphic symbols for the elements, but these proved difficult for chemists and their printers.
The British Library 8906.b.18.

his intellect to many public and social problems, in manufacture, agriculture and hygiene, and his tragic and unnecessary death at the hands of the revolutionaries was the result of the unfortunate associations which he had with the *ancien régime*.

But Lavoisier was not able to elucidate how and why certain elements combine while others do not; at what level does the combination occur, and why does simply exposing two elements to each other not necessarily produce a reaction? The most important single step on the road to answering these questions came when John Dalton (1766–1844) argued in his *New System of Chemical Philosophy*, 1808, that these events occurred at the atomic level. The atom as a concept was an ancient one, but was no more than a theoretical possibility, and it had played no part in Aristotelian science. An interest in atomic theory revived during the seven-

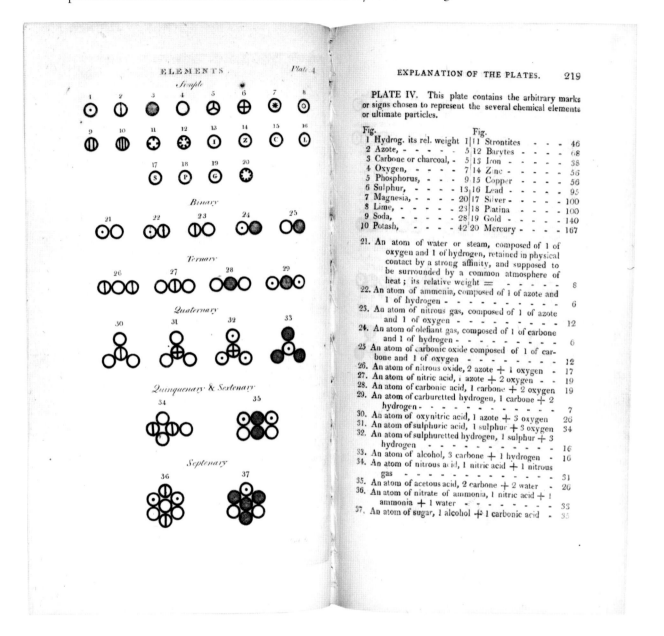

teenth century, largely through the discovery that air has weight and qualities, which could only be explained by supposing that it consisted of particles so minute as to be intangible and invisible, yet aggregating to a definite physical mass. Boyle subscribed to this view, and extended it to solids and liquids too. It was very influentially stated by Newton: 'God in the beginning formed matter in solid, massy, hard, impenetrable, moveable particles … even so very hard as never to wear or break in pieces; no ordinary power being able to divide what God him-self made one in the first creation'. How this might relate to chemical combination had been hinted at by Lavoisier's contemporary, Joseph-Louis

Proust in 1794, when he published his law of definite proportions. Proust found that when he analysed known compounds into their constituent parts, they always contained the same proportions of their elements by weight, for example copper carbonate was always five parts copper, four parts oxygen and one part carbon. Dalton pondered this, and at last came to the conclusion that this was explicable if combination occurred at the atomic level: minute particles of different substances, having different weights, combined one to one to produce a compound. Of course Dalton never saw or handled atoms directly, but they provided a unifying principle through which to understand chemical reaction, and he was able to calculate the weights of twenty elements, using the lightest, hydrogen, as base 1. Thus the element was first given a mathematical definition, for the atom of each element was unique in its weight. On this basis, the weight of oxygen was 7, iron 50, mercury the heaviest at 167, and so on. Dalton knew that different compounds of the same elements result from their combining in different proportions, for example carbon dioxide contains twice as much oxygen by weight as carbon monoxide. Unfortunately Dalton confused the terminology, for he spoke of the combination as itself an atom, where we should call it a molecule. He proposed a system of graphic symbols to represent the elements, but these were visually rather unclear, and would have been very difficult to follow when used in chemical equations as they later developed. Despite certain inconsistencies, Dalton's theory was a stroke of genius, which would later be vindicated by physical experiment, and which provided the fundamental, but still baffling, foundation of the modern understanding of matter.

No greater contrast could be imagined between the careers of Dalton, the retiring, pious dissenter, and Sir Humphrey Davy who became rich, showered with honours, and a London celebrity. His earliest ambitions were poetic, but he progressed to science, and, as he later said in a revealing phrase, 'my visions fled before the voice of truth'. He first achieved social success by demonstrating the anaesthetic – and hysterical – effects of nitrous oxide, or laughing gas, and he became a public lecturer before large, fee-paying audiences. He turned to electrical experiments, and reasoned that if chemical reactions could produce an electric current, as Volta had shown, then electric current should stimulate chemical reactions. He was the first to use electrolysis in the Voltaic battery to isolate sodium and potassium, elements so reactive that they do not occur in their pure form in nature. Davy's invention of the safety lamp for use in mines was based on an ingenious separation of the heat of a flame from its light, by means of a metal gauze; this achievement made him perhaps the first public hero in scientific history. Davy cannot be credited with great theoretical discoveries in the manner of Dalton, but he pioneered the borderland between chemistry and physics which his pupil, Faraday, would exploit so brilliantly.

Portrait of John Dalton
Science Museum, London

On the foundations laid by Dalton, vital theoretical advances were made by European chemists between 1810 and 1850. In post-revolutionary France, science had begun to be incorporated into higher education in the form of the *École polytechnique* in Paris and an intellectual tradition was founded, while in England the private amateur was still the norm. Joseph-Louis Gay-Lussac (1778–1850) was first a student then a teacher at the *École,* where he made experiments which showed that the volumetric proportions in gas mixtures was important, not the weights as Dalton had believed. Thus water was two volumes of hydrogen to one of oxygen, and this volumetric approach paved the way for the chemical language we now use, in which water is H_2O, and not HO as it would have been to Dalton. It was the Italian chemist Amedeo Avogadro (1776–1856) who linked the atomic theory with the volumetric, by postulating that the atoms in certain elements, and the atoms in all compounds, exist joined in an atomic *group*, which he called the molecule (Latin = 'little mass'). Thus oxygen is O_2, and thus only half a volume of oxygen is needed to combine with one volume of carbon monoxide to form one volume of carbon dioxide, since each atom of oxygen joins one molecule of carbon monoxide. It was these molecules which were capable of being separated then re-formed in new patterns during chemical reactions, while the individual atoms remained unchanged, and this was the true basis of what had been perceived as the conservation of matter. This was a crucial step, but it was not widely accepted because in practice these early chemists knew in very few cases how many atoms were in each molecule. The Swede, Jöns Berzelius (1779–1848), advanced chemical notation enormously by adopting many of the alphabetic

Exempel af några Dubbelsalters sammansättning.

Namn.	Formel.	Partikelns vigt.	Starkste basis	Svagare basis ell. fyra.	Starkaste fyran.	Vatten.
Carbonas Magnesico Calcicus, Bitterspat, Dolomit	$Ca\ddot{C}^2 + Mg\ddot{C}^2$	2330,10	30,56	22,18	47,26	
Fluosilicias Ammonicus	$(3NH^6+2Si)+(3NH^6+3F)$	3310,57	32,88	36,13	24,99	
Hydricus	$3\ddot{F}Aq^9+2Si^3F^2$	5540,55		43,06	44,67	12,27
Kalicus	$3\ddot{K}F+2Si^2F^3$	8400,44	42,13	28,41	29,46	
Oxalas Ammonico-Cupricus	$2NH^6\ddot{O}Aq^2+Cu\ddot{O}^2Aq^2$	3908,95	11,02	25,36	46,23	17,39
biammonico Cupricus	$(3NH^6+2O)+2Cu^3\ddot{O}^2+Aq$	9418,55	6,86	63,16	22,78	1,20
triammonico-Cupricus	$2(3NH^6+\ddot{O}Aq^3)+Cu^3\ddot{O}Aq^6$	7433,09	17,39	40,01	24,31	18,29
Oxalas Kalico-Cupricus, Var. 1:ma	$\ddot{K}\ddot{O}^2Aq+Cu\ddot{O}^2Aq$	4204,83	28,05	23,58	42,98	5,39
Var. 2:da	$\ddot{K}\ddot{O}^2Aq^2+Cu\ddot{O}^2Aq^2$	4431,37	26,63	22,37	40,78	10,22
Natrico-Cupricus	$Na\ddot{O}^2Aq+Cu\ddot{O}^2Aq$	3806,84	20,54	26,04	47,47	5,95
Murias Ammonico-ferrosus	$2NH^6\ddot{M}+Fe\ddot{M}^2$	3376,19	12,76	26,02	61,22	
Hydrargyricus	$2NH^6\ddot{M}Aq+Hg\ddot{M}^2$	5455,89	7,90	50,07	37,88	4,15
Platinicus	$2NH^6\ddot{M}Aq+Pt\ddot{M}^2$	4139,52	10,41	34,19	49,93	5,47
Kalico-platinicus	$\ddot{K}\ddot{M}^2+Pt\ddot{M}^2$	4661,94	25,30	30,36	44,34	
Natrico-Platinicus	$N\ddot{M}^2+Pt\ddot{M}^2$	4263,95	18,34	33,19	48,47	
Hydro-Carbonas Cupricus	$Cu Aq^2+2Cu\ddot{C}^2$	4302,02		69,13	25,60	5,27
Magnesicus (v. Magnesia Alba).	$Mg Aq^2+3Mg\ddot{C}^2$	4625,00	44,69		35,72	19,59
Zincicus	$Zn Aq^6+3ZnC$	5531,39	72,78		14,93	12,29
Muriocarbonas Plumbicus	$Pb\ddot{M}^2+Pb\ddot{C}^2$	6813,96	81,86	8,08	10,06	
Sulphas Aluminico-Ammonicus	$NH^6\ddot{S}+Al\ddot{S}^3$	2862,40	7,53	22,44	70,03	
Kalicus (Alumen).	$\ddot{K}\ddot{S}^2+2Al\ddot{S}^3$	6473,75	18,23	19,84	61,93	
c. Aqua	$\ddot{K}\ddot{S}^2+2Al\ddot{S}^3+48Aq.$	11910,60	10,15	10,54	33,66	45,65
Natricus	$Na\ddot{S}^2+2Al\ddot{S}^3$	10085,04	7,75	12,74	79,51	
Ammonico-Cupricus	$2NH^6\ddot{S}Aq^2+Cu\ddot{S}^2Aq^{10}$	7017,30	6,15	14,13	57,13	22,59
triammonico-cupricus (Cuprum Ammoniacum).	$4(3NH^6+\ddot{S})+Cu^3\ddot{S}^2Aq^6$	9246,01	27,96	32,17	32,52	7,35
Ammonico Kalicus	$\ddot{K}\ddot{S}^2+2NH^6\ddot{S}Aq^2$	6073,06	19,43	7,09	66,02	7,46
Magnesicus	$2NH^6\ddot{S}Aq^2+Mg\ddot{S}^2Aq^{10}$	6542,63	6,59	7,90	61,28	24,23
Calcico-Natricus (Glauberit).	$Na\ddot{S}^2+Ca\ddot{S}^2$	5503,18	14,21	12,94	72,85	

X

The new chemical language, 1818: among the innovations of the great Swedish chemist Jöns Berzelius, was the system of symbols which we now use – Ca for calcium, Mg for magnesium and so on. These proved far simpler than Dalton's symbols, and far better adapted to chemical formulae.
The British Library 1144.k.4.

symbols which we now use – O, H, Fe, Au, and so on. These proved universally acceptable where Dalton's had not, and facilitated the use of equations to describe chemical reactions. Yet Berzelius could still speak of an atom of alcohol, as well as an atom of hydrogen, while others could analyse water volumetrically as one atom of hydrogen and half an atom of oxygen. It was Berzelius who invented a number of fundamental chemical terms, such as catalyst and protein.

In mid-century the language of chemistry was still in considerable flux, even confusion, nor were all chemists, and certainly not all physicists, convinced of the physical reality of atoms; at that stage they were still a convention, a working hypothesis. It became necessary for the identity of chemistry – if it was to be taught as a discipline in colleges for example – that its language should reflect a consistent understanding of its processes. In 1860 the first international chemical congress took place in Karlsruhe, at which Stanislao Cannizzaro (1826–1910) successfully urged the adoption of Avogadro's ideas of atoms, composite atoms and molecules, all to be analysed volumetrically. The results of this conference were the culmination of one distinct phase of the chemical revolution which Lavoisier had inaugurated. The question of the physical nature of atoms and molecules, and the forces which bound them together were no less baffling, but came to be seen as the province of physics, while chemistry set out to systematize the thousands of possible reactions and compounds arising from the fifty or so known elements – a number which in fact was steadily being augmented as analytical techniques improved. The most dramatic technical advance was spectroscopy, which had been initiated as early as 1814 by Joseph Fraunhofer, but which became of major importance in the hands of Gustav Kirchhoff and Robert Bunsen in 1859, when they showed that different substances emit different light spectra as they burn, as distinct as a system of fingerprints, which might be recorded and compared. One result of these techniques was that the atomic weights of more and more elements were known, and it was felt that the weight was the one objective fact which could be stated about the atom, and that it must

embody some important feature of each element. Dmitri Mendeleyev (1834–1907) felt very acutely as a professional teacher of chemistry that he should be able to classify the elements into some significant order, and indicate material relationships between them. In 1869 while working on the composition of his textbook *Principles of Chemistry* (published in English translation in 1891), he discovered that when the elements were tabulated in ascending order of their weights, they fell naturally into groups which shared important characteristics. Thus the heavy metals gold, mercury and lead are adjacent, while the alkali metals which do not occur in nature, calcium and potassium, fall together, as do the light, corrosion-free metals titanium, vanadium, and chromium. This 'periodic table' thus became an immensely practical organizing tool in which theory and practice reinforced each other. Its validity was amply confirmed when certain gaps in it were filled by the discovery of elements whose existence Mendeleyev predicted because there were gaps in the numerical sequence. It is now known that it is the atomic number (that is the number of protons and electrons) rather than the weight that is significant, but this data was not available to Mendeleyev, and most of his weight-relationships are still valid. In the atomic age, the number-relationship would become still more important as charting the transformation of one element into another as electrons were gained

Portrait of
Dmitri Mendeleyev
Science Museum, London

or lost. Mendeleyev's achievement was perhaps the high-point of the remarkable nineteenth-century investigation of the atom, which could not be seen or examined directly, but whose nature was so tenaciously probed by reason alone.

CLASSICAL PHYSICS

The science which we now call physics acquired its modern form between the years 1800 and 1870, when certain vague, hypothetical entities such as electric fluid, caloric and ether, were replaced or re-defined as measurable forces, and when the concept of energy emerged as a unifying principle. This new understanding had immense practical consequences in terms of power generation and measurement, and a second industrial revolution began in which it may truly be said that science transformed society. Even before 1800, Thomas Young (1773–1829) had begun to question the accepted view of light as a stream of particles, and to revive the wave theory of Huygens, which better explained many aspects of the behaviour of light. In France, Augustin Fresnel (1788–1827) reached the same conclusion independently through a study of polarization. Apparently of theoretical interest only, the Young-Fresnel wave theory of light was to act as a revelation when James Clerk Maxwell integrated it with his theory of electromagnetic radiation in the 1860s.

A comprehensive theory of heat was soon found to be of paramount importance, and its study was clearly stimulated by industrial processes. Even before the end of the eighteenth century, the American engineer Benjamin Thompson (1753–1814), better known as Count Rumford, had been sceptical of the existence of 'caloric' as a fluid element. He noted that in the boring of cannon barrels, heat was produced apparently from nowhere, and that the most careful measurements failed to reveal the slightest alteration in the weight of a given volume of water as it froze or melted or was heated. Heat as a substance could therefore be discounted, and he was inclined to believe that it arose from vibratory motion. The man who formulated the most original theory of heat was the Frenchman Sadi Carnot (1796–1831) who unfortunately died young before his work was complete or widely disseminated. The interesting thing about Carnot's approach was that it was based on careful study of the functioning of the steam engine. Carnot pointed out that heat moved from source to component in the engine, that it was dissipated as it moved, and that work was done as it was dissipated. The mechanical work performed by the engine resulted from the transfer of heat from source to component. Heat could be seen as equivalent to motive power, and since it was also known that motive power produced heat, in machines and animals, he inferred that the heat in any system was finite and constant, merely changing its form.

A similar approach to Carnot's was taken independently in England and in Germany by a number of leading researchers in the 1840s, whose work converged rapidly into the mature doctrine of energy. James Joule (1818–1889) designed a series of experiments to measure precisely how much work was needed to perform simple mechanical tasks such as turning a wheel, and conversely how much heat was generated in and around the mechanism. The result was always an equivalence in heat and work between input and output. Taking a simple unit of heat – that needed to raise the temperature of one pound of water by one Fahrenheit degree – and a simple unit of work – that needed to raise one pound weight one foot – he discovered that the 'mechanical equivalent of heat' was 838 'foot-pounds'. But this was not all, because Joule found the same equivalence in chemistry and electricity: if zinc is plunged into concentrated acid it dissolves quickly and gives off heat; but if the zinc is made an element in a battery, a current passes slowly from it, which will heat a wire; if in turn the current is made to drive an electric motor, the heat in the wire is less. In each case the amount of heat varies in proportion to the work done. Joule called this principle 'living force', stating that 'wherever living force is apparently destroyed … an exact equivalent of heat is restored; the converse is also true, namely that heat cannot be lessened or absorbed without the production of living force … in these conversions, nothing is ever lost'. This is easily recognizable as the first law of thermodynamics, and it was stated explicitly by Rudolf Clausius (1822–1888): in a closed system the total amount of energy is constant. The second law followed, namely that heat must always flow from a hotter object to a colder one, and never the reverse, and Clausius foresaw that the end of this process on the cosmic scale would be the 'heat-death of the universe'. The word 'energy' (Greek = 'work') had been briefly introduced before, but it was William Thomson, Lord Kelvin, (1824–1907) who used it to unify ideas about heat and its transformations, and who formulated the third law of thermodynamics, that it is impossible to go on taking heat from an object without limit, because if heat is to flow, there must always be a temperature difference between an object and its environment. This becomes impossible at the absolute zero at -273 degrees centigrade, and this figure provided an essential fixed scale with which energy and heat could be measured, as distinct from temperature.

All these findings were summed up in the phrase 'conservation of energy', which was the title of a work published in 1847 by Hermann Helmholtz

(1821–1894), who extended the model of the machine to the universe as a whole, in which the sum of all energies remains constant through repeated transformations. Any measurable form of physical activity was convertible to any other form – heat, work, electricity, chemical reaction – and these forces remain finite and unchanging. The parallel with the chemists' discovery of the conservation of matter through the change of forms during chemical reactions, is striking. Helmholtz also applied this model to animal physiology: that muscular action involves the oxidization of sugar, producing energy. With the contemporary knowledge of photosynthesis, the realms of animal and plant life, chemical and solar energy were being bound ever closer in an interdependent system. Helmholtz's physiological theory was paralleled by that of his brilliant contemporary, Julius Mayer (1814–1878), indeed Helmholtz generously acknowledged that Mayer had truly understood the principle of energy-conservation before any one else. Newton's theory of gravitation had explained something central about the large-scale structure of the universe, while energy-theory now revealed something new about its inmost workings. Heat had become a dynamic force, where the old

Portrait of
Hermann Helmholtz
Science Museum, London

caloric theory had regarded it as a static substance. Helmholtz built this insight into an intriguing model of scientific description: if we imagine matter dispersed into its elementary particles – whatever they may be – then all conceivable change lies in the spatial re-arrangement of those elements. All the phenomena which science investigates are therefore reducible to the forces that produce those changes, and science consists in the identification and measurement of those forces.

New discoveries about electricity in the 1820s lifted it far beyond the realm of the parlour game, and reinforced the trend towards a unified or interrelated view of physical forces. George Ohm (1787–1854) used Coulomb's torsion balance to investigate what was happening when a current passed through a wire. He conceived that the current moved in a flow as if passing from particle to particle, and that the wire medium offered a resistance which absorbed some of the current and manifested itself as heat. He used the phrase 'electroscopic force', which we

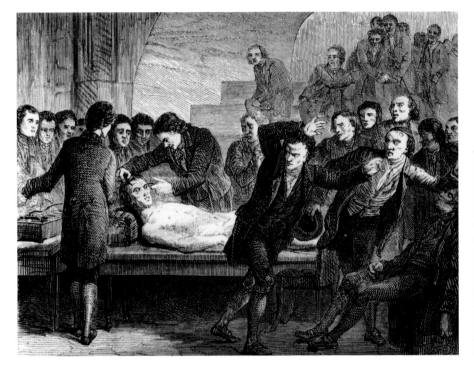

The influence of Frankenstein? In 1818 Dr.Andrew Ure conducted experiments in Glasgow in which the nerves of a corpse were stimulated by electricity, producing sufficient reaction to sustain the belief that electricity might indeed be the primal force of life itself. From Figuier's *Merveilles de la science*, 1867.
The British Library 8765.g.15(1)

now call electromotive force, and drew an analogy between heat flowing from a hotter to a cooler substance and the 'potential difference' between two electrical terminals. Ohm's law of electromotive force and resistance enlarged the vocabulary of electrical science, and would later indicate how electricity could be channelled to produce heat and light. In 1820 the Dane, Hans Oersted (1777–1851), was the first to demonstrate the vital connection between electricity and magnetism: in his experiments in 1819 an electric current in a wire was made to deflect a compass needle. André Ampère (1775–1836) showed the converse, that a wire coil behaves like a magnet when a current is passed through it, and that two parallel wires carrying currents will attract each other if the current flows in the same direction, but repel each other when it flows in the opposite direction. The earliest form of 'galvanometer', with which an electric current is measured, made use of this property of deflecting a small magnetic needle.

Electricity was still an empirical, non-theoretical pursuit, so that Michael Faraday (1791–1867) could, with no formal education but with the benefit of a few years' thoughtful experimentation, find himself at the frontier of electrical science. In his crucial demonstration of 1821, Faraday constructed an electric circuit containing a freely-hanging magnet suspended above a terminal, and freely-hanging terminal above a magnet. When the circuit was completed, both moving parts rotated: this was the first, historic, conversion of electrical energy into mechanical energy. Further experiments showed that magnets alone could 'induce' electricity, and that there could be no doubt that these were but two aspects of the same force. Faraday's experiments became the models on which the first electric motors and dynamos were built. Faraday brilliantly rationalized what was happening in these experiments – how these forces were operating across empty space – by producing the theory of the magnetic or electrical field, whose lines of force could be

Painting of Michael Faraday in his laboratory at the Royal Institution *c*.1840. Without formal education and ignorant of mathematics, Faraday possessed outstanding intuition, and laid the foundations of electrical science.

The Royal Institution/Bridgeman Art Library

graphically revealed by the now-familiar pattern of iron filings around a magnet. Faraday speculated that electricity functioned by changing the patterns in the elementary particles of matter; this approximates quite well to the modern understanding of electricity as a flow of electron jumps.

Faraday was not able to give his work any final theoretical or mathematical expression, and this achievement fell to James Clerk Maxwell (1831–1879) who in several works culminating in his *Treatise on Electricity and Magnetism,* 1873, sought to show 'how by a strict application of the ideas and methods of Faraday, the connection between the very different orders of phenomena which he discovered may be placed before the mathematical mind'. Essentially, Maxwell constructed a mechanical model to explain how electrical or magnetic forces propagate themselves. The strength and distribution in space of electric or magnetic fields were analysed and related to each other. His general conclusion was

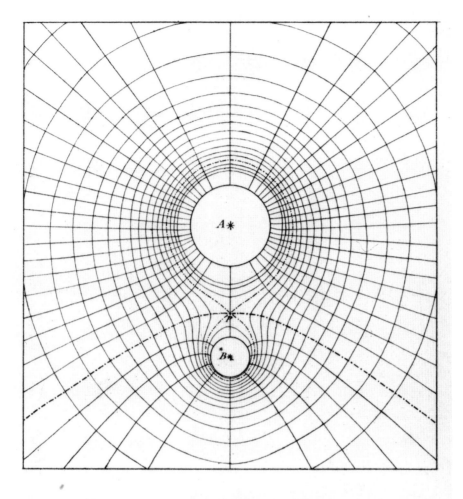

Lines of Force and Equipotential Surfaces.

$A = 20$. $B = 5$. P, *Point of Equilibrium*. $AP = \frac{2}{3} AB$.

Electromagnetic fields visualized by James Clerk Maxwell. Maxwell gave mathematical form to Faraday's intuitive discovery that electricity and magnetism create fields of force. This diagram illustrates the mutual repulsion of two unequal fields around two electrodes. From Maxwell's *Treatise on Electricity and Magnetism*, 1873.

The British Library 12205.s.1.

that they travelled in high-velocity waves, in the same manner as light was now believed to do in the Young-Fresnel wave theory. On calculating the characteristics and velocity of these waves, Maxwell found himself unable to resist the inference 'that light consists in the transverse undulations of the same medium which is the cause of electric and magnetic phenomena'. Maxwell had discovered the concept of electromagnetic radiation, and he predicted correctly that other forces with different wavelengths would be discovered within a 'spectrum'. As early as 1800 William Herschel had noticed, while studying the spectrum of sunlight, that heat was clearly measurable beyond the red end of the spectrum, and correctly inferred that radiation invisible to the eye, later known as infra-red rays, was responsible. By 1868 Anders Angstrom (1814–1874) had devised a way of measuring the wavelengths of both visible and invisible waves; his unit of one ten-millionth of a millimetre was later called the Angstrom, the smallest unit of measurement in scientific use.

Helmholtz began to study the problem of finding experimental proof of Maxwell's model and in a series of experiments in Karlsruhe, one of his pupils,

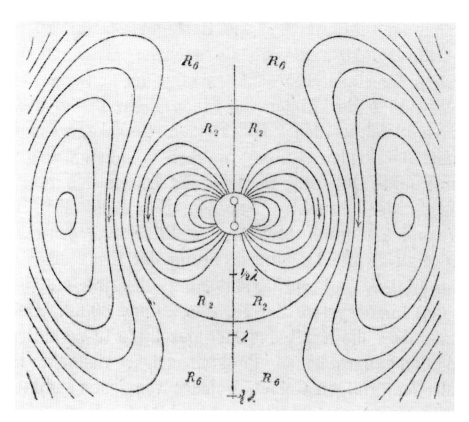

Mapping electromagnetic waves by Heinrich Hertz. Hertz devised experiments to verify Maxwell's theory that electricity was propogated as waves in space, like light. When Marconi adapted Hertz's experiments in the 1890s, he named this wave effect 'radio'. From Hertz's *Ausbreitung der elektrischen Kraft*, 1892.

The British Library 8757.h.24.

Heinrich Hertz (1857–1894), succeeded in producing electromagnetic waves from an unclosed circuit, and in detecting them across a space with an unclosed loop of wire. He refracted electric waves through a pitch prism, just as light is refracted, and condensed them with concave mirrors, and he measured the length of the electric waves: from their length we now know that they were radio waves, and their speed was identical to that of the speed of light. The essential unity of the forces of nature was steadily being revealed in physics as in chemistry.

The medium which Maxwell and Hertz believed was necessary to embody these waves was still the ether, the most tenacious of the hypothetical entities of pre-atomic physics. The nature, and even the reality, of the ether was indeed one of the crucial mysteries of nineteenth-century science. If light, electricity and magnetism were, Maxwell's phrase, "undulations in a medium", then that medium must permeate the universe. It must be so tough that it could bear the enormous stresses implied by gravity, and yet so diaphanous that both electromagnetic forces and heavenly bodies might whirl through at unimaginable speeds; moreover this baffling substance was completely undetectable by any physical test. The leading physicists of the age – Kelvin, Maxwell, Helmholtz and others –

considered that all forces were the result of material in motion, and hence felt compelled to uphold the doctrine of the ether. In the 1880s A.A. Michelson (1852–1931) and E.W. Morley (1838–1923) performed a series of intricate experiments in America which aimed to establish the different speeds of light in the direction of the earth's path, and at right-angles to it. The result of what has been called the most important negative experiment in the history of science was that no difference was found. This in itself raised a most puzzling question about the speed of light, but furthermore it seemed to prove that there is no motion of the earth relative to the ether. By the 1890s there was a growing conviction that, while the ether was a philosophical necessity, nature behaved as though it did not exist. Part of its philosophical appeal lay in the concept of absolute space. In the pre-Copernican universe the earth was believed to be at rest, central in the sur- rounding cosmos. Then the Sun was considered motionless and central, until Edmond Halley and others demonstrated that the stars, including the Sun, were actually in motion. In the Newtonian universe of absolute space therefore, rest and motion must be defined in respect of the ether. The doubts about the ether which were gathering at the end of the nineteenth century would culminate in Einstein's rejection of the concept of absolute space. This provided the philosoph- ical thrust, which, when added to the experimental result of Michelson and Morley, led to the final downfall of the idea of the ether.

ASTRONOMY

The characteristic of nineteenth-century astronomy was the widening of its hori- zons from the traditional study of planetary motion to stellar astronomy, thus beginning a new phase in man's understanding of the scale and nature of the uni- verse. Such a revolution had been implicit since the sixteenth century with its insight that the Sun is a star like the other stars, but the difficulties in measuring or deducing facts about the stellar realm were enormous, and were not greatly helped even by the early telescopes. This situation began to change when the technique of making achromatic lenses was mastered in the 1750s by John Dol- lond and others. One of the first significant observational programmes of the outer heavens was that of Charles Messier (1730–1817) who turned his attention to the nebulae, the cloud-like or misty stars which were something of a nuisance to eighteenth-century astronomers, especially those studying comets. By 1784 Messier had charted and numbered 101 nebulae, (which still bear their M-num- bers), without of course being able to elucidate their true nature. Some were in fact clouds of interstellar gas, some clusters of stars, while others are what we now know to be galaxies beyond our own; determining the nature of Messier's nebu- lae was to become a central part of modern cosmology.

The first great stellar astronomer was undoubtedly William Herschel (1738–1822), although rather ironically he first achieved fame as the discoverer of the planet Uranus in 1781, the first addition to the solar system in four thou- sand years of astronomical history. Herschel had built for himself the most powerful telescopes of his time, culminating in a huge 40-foot reflector com- pleted in 1789; the story is told that on the second night of its operation he discovered a new satellite of Saturn (later named Enceladus), and, having an unrelated paper going through the press, hastily added the famous words 'P.S. Saturn has six satellites'. Herschel's aims lay not among the planets however, but among the stars, for he had conceived the ambitious plan of mapping their distri- bution through space. All conventional star-charts, and all positional astronomy,

Portrait of
William Herschel
Science Museum, London

William Herschel's drawing of the nebulae he had observed. Having built the most powerful telescopes of his era, Herschel charted star positions in an attempt to determine the large-scale structure of the universe; in doing so, he noticed and recorded the shape of the nebulae, whose interpretation was to become central to modern cosmology. From *Philosophical Transactions of the Royal Society,* 1811. The British Library C.144.l.3.

treat the heavens as a two-dimensional plane in which objects are related by angular distance. This was part of Herschel's plan, but he was also attempting something far more original, namely to survey the heavens in three dimensions, to determine distances that were radial from the earth, and hence to offer a picture of the structure of the cosmos in depth. Of course Herschel had no data whatever as to stellar distances, but he made the basic assumption that all stars are more or less of equal absolute brightness, and that the luminosity we perceive is an index of distance. He then proceeded to count stars, thousands and thousands of them in small, set areas of the sky. He reached the general conclusion that the stars were grouped together in a vast disc formation, thus explaining the concentration of stars in the Milky Way as the edge-on view of the disc; this was correct, and despite the fact that Herschel naturally supposed that all the objects in the sky were part of the one system, his thirty-year programme to determine the structure of the universe marks the beginning of observational cosmology. In 1845 the Earl of Rosse used his massive reflecting telescope with its six-foot mirror to resolve the nebula M51 into a spiral galaxy, an object whose nature and distance from the earth could not be reconciled with the Herschelian structure.

New standards in technical instruments, notably the transit-instrument and the micrometer, paralleled the advances in the telescope, and they made possible the

The discovery of the planet Neptune in 1846 became the subject of a fierce priority dispute between the French supporters of Urbain Leverrier and those of John Couch Adams in England. In this French cartoon, Adams peers merely at Leverrier's announcement of the discovery, while the Frenchman sees the planet itself. From *L'Illustration*, 8, 1846.

The British Library, Newspaper Library, Colindale

resolution of a centuries-old problem – the measurement of stellar parallax. It was Friedrich Bessel (1784–1846) at Königsberg, who in 1838, after observing a relatively close star in the constellation Cygnus for more than a year, found a movement of less than one third of a second of arc. Bessel had at last found empirical proof that the earth orbits the Sun, but he did more, for he went on to calculate from its parallax-shift its distance from the earth, which he determined to be 657,000 times the distance from the Earth to the Sun (in modern terms 10.3 light-years). Perhaps Bessel's achievement should be counterpointed by that of Jean Foucault whose pendulum experiment of 1851 at last gave proof that the earth rotates upon its axis: the path of a long pendulum free to swing in any direction will be seen to change its angle, but in fact it is the earth which is moving beneath the pendulum.

The greatest impact of any new instrument in astronomy was made by the spectroscope, which can identify materials as they burn by analysing their light. It was developed as early as 1814 by the Munich instrument-maker Joseph Fraunhofer, who noticed that the spectrum was intersected by numerous black lines, which he was unable to explain, although he saw that the distribution of the lines varied with different light-sources. The chemical significance of the spectroscope was grasped quite early, as the pioneer of photography, Fox Talbot, wrote in 1826 'a glance at the prismatic spectrum of a flame may show it to contain substances which it would otherwise require a laborious chemical analysis to detect'. But its potential for analysing starlight was first realized during the years 1859–1862 by Gustav Kirchhoff and Robert Bunsen, who established that since each element produced its own distinctive spectrum, sunlight would indicate which chemicals were present in the stars. The importance of this discovery for physics and astron-

Foucault's pendulum in the Pantheon, Paris, 1851. A pendulum swings in a constant plane, and Foucault applied this principle to prove, three centuries after Copernicus, that the earth rotates. To observers, the pendulum appeared to turn through 360 degrees in 24 hours, but in fact it was the earth rotating beneath it.

From *L'Illustration,* 17, 1851.

The British Library, Newspaper Library, Colindale

omy can hardly be overstated, for it had long been imagined that the stars would remain forever out of reach of human study. Within a few years Norman Lockyer in England had identified an unknown element in the Sun, which he called helium, and spectroscopic analysis went on to provide some of the fundamental data of modern cosmology. Perhaps the most important was the discovery by William Huggins (1824–1910) that the light from the stars shows them to be moving away from the earth. This was an application of the principle explained by Christian Doppler in 1842, that the wavelength of radiation from an energy source becomes longer if the source is receding. When applied to starlight, this meant that the long- wavelength section, the red part, of the spectrum is stretched or magnified in the so-called 'red shift', while the blue is diminished. In 1868 Huggins found that the star Sirius is receding from the Earth at a speed of almost thirty miles per second. The technique of measuring the red shift was later to assume the utmost importance when it was applied to galaxies outside our own.

Nineteenth-century stellar astronomy can be seen in fact as an unfolding series of such clues or stepping-stones leading towards a new cosmology. Another was the investigation of the so-called 'Olbers paradox', that since the heavens are filled with stars scattered throughout space, why is the night sky dark, why does not the eye always encounter a star along every line of sight? In 1861 Johann Mädler divined part of the answer, that the light had simply not had time to reach the Earth; for this to be true, the time of the light's travel must clearly be less than the age of the universe, and given the known speed of light, this had profound implications. Approximate values for the speed of light had been current since the late seventeenth century, but in 1849 Armand Fizeau in Paris devised an ingenious optical mechanism which gave him a new and more accurate speed of 195,000 miles per second, just 5% higher than the true figure. The instrumental revolution continued with the application of photography to the heavens. The ability to preserve photographic images of the sky for later analysis was especially important in the new study of nebulae, where there was formerly uncertainty about what had actually been seen. In the 1880s the amateur Isaac Roberts in England secured some historic first images of the nebulae, and then George Hale in America began to produce his spectacular series of astro-photographs which contributed so much to twentieth-century cosmology.

Yet the more traditional pursuit of planetary astronomy could still produce its own revolutionary discoveries. In the 1820s the orbit of the new planet Uranus was proving puzzling to astronomers, for it displayed irregularities which suggested either that Newton's laws of gravitation did not strictly apply at such great distances from the Sun, or that that a still further planet beyond it was disturbing its expected path. Many astronomers began the task of predicting the probable position of such a body, giving rise to one of the closest-run priority disputes in scientific history. Urbain Leverrier (1811–1877) submitted detailed predictions of the planet's position to the Berlin Observatory in September 1846, and within days it was found. The international celebrations became confused when it was made public that an English mathematician, John Couch Adams (1819–1892), had reached the same result exactly one year earlier. Adams had sent his his work to the Professor of Astronomy at Cambridge and the Astronomer Royal, but they had neglected to verify it. There was a curious sequel to this episode when Leverrier, puzzled by certain discrepancies in the orbit of Mercury, predicted that yet another planet must lie inside its orbit; he provisionally named it Vulcan, but it was of course never found. The finding of Neptune was another vindication of Newtonian dynamics, and it showed how productive this form of mathematical astronomy could still be. Before the Neptune episode however there had been

another addition to the solar system, whose means of discovery was by no means so rigorous. In 1772 two German mathematicians, Johann Titius and Johann Bode, had drawn attention to a curious sequence in the planets' distances from the Sun. The sequence runs:

$$0+4=4 \qquad 3+4=7 \qquad 6+4=10 \qquad 12+4=16$$
$$24+4=28 \qquad 48+4=52 \qquad 96+4=100 \qquad 192+4=196$$

If the totals in each case are divided by 10, the result is very close to the distance of each planet from the Sun, given in astronomical units, that is the earth's distance from the Sun:

$$\text{Mercury } 0.4 \quad \text{Venus } 0.7 \quad \text{Earth } 1.0 \quad \text{Mars } 1.6$$
$$\text{Jupiter } 5.2 \quad \text{Saturn } 9.5 \quad \text{Uranus } 19.2$$

The fit is generally good, but there was evidently a gap between Mars and Jupiter, and it was naturally asked whether an undiscovered planet could possibly be orbiting the Sun at 2.8 astronomical units distance? In 1801 the Italian astronomer Giuseppe Piazzi did indeed locate a planetary body in the predicted orbit, which he named Ceres. Yet it proved to be very small (it has a radius of 457 kilometres) and the situation became more puzzling when, in the following few years other still smaller bodies were found to share the same orbit. By 1872 more than 100 'asteroids' had been located, and it was long believed that they formed the remains of a large planet which had somehow disintegrated. The status of the Bode-Titius law is still unclear: Neptune does not fit the pattern well, with Pluto it breaks down entirely, and it appears to be an accidental relationship, in the manner of Kepler's geometric model of the solar system. At the end of the nineteenth century the solar system was still not fully discovered, no totally convincing theory had emerged concerning its origin, or the question of its uniqueness. But it is true to say that the frontiers of astronomy had by then moved far beyond the solar system, and data was accumulating which would lead to a radical redefinition of the scale and structure of the universe.

GEOLOGY AND EVOLUTION

The physical sciences of the nineteenth century found new techniques to probe the structure of matter, and a new language in which to express their discoveries. That language was becoming increasingly specialized, so that probably no more than a few hundred scholars in all Europe could personally evaluate Maxwell's equations or Mendeleyev's atomic tables. Yet the fruits of the new science were soon visible everywhere in the new techniques which were producing synthetic substances and electrical power. However the quintessential science of the age proved to be that blend of geology and biology which is summed up in the concept of evolution. Evolution reflected and reinforced something absolutely central in nineteenth-century culture, namely the belief in progress – progress defined both physically and morally. Evolution was non-technical, it could be understood by anyone, and it re-defined man as, in the literal sense, a part of nature. It concerned man himself more intimately than any discoveries in physics or chemistry, and it was a coming-of-age, a final dissolution of mythical or divine explanations of his origins. The theory of evolution – Darwinism – was officially born in 1859 with the publication of *The Origin of Species,* an event as significant as the publication of the great books of Copernicus or Newton. Yet the striking thing is just how long the concept of biological transmutation had been develop-

The Orion Nebula photographed in 1889 by Isaac Roberts. Roberts was one of the pioneers of astro-photography, which, together with spectro-scopy, revolutionized stellar astronomy in the later nineteenth century. From Roberts' *Selection of Photographs of Stars and Star-Clusters,* 1893.
The British Library 8563.i.24.

ing before Darwin, and the length of that process makes it all the more important to establish what his unique contribution was.

Curiously, one of the important transitional figures was the great naturalist's own grandfather, Erasmus Darwin (1731–1802), physician, botanist and inventor.

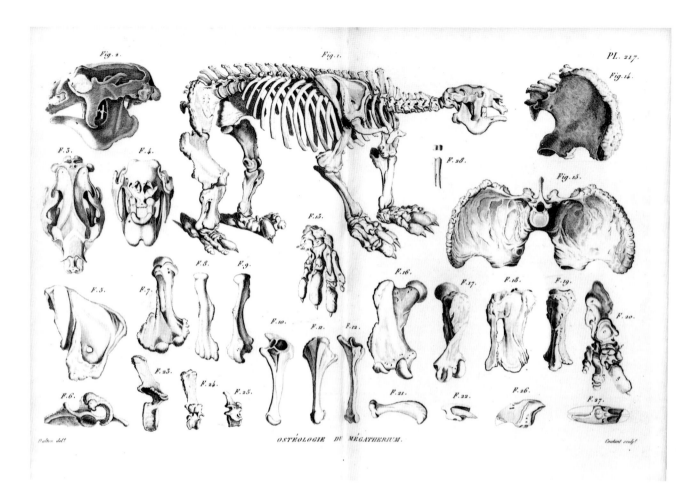

OSTÉOLOGIE DU MÉGATHERIUM.

Cuvier's reconstruction of the skeleton of the giant sloth, or *megatherium*. Cuvier, the foremost palaeontologist of his time, was however an anti-evolutionist. He believed that species had become extinct suddenly in catastrophic events, and that there was no link between living and dead species. From his *Recherches sur les Ossements Fossiles*, 1825

The British Library 462.i.10-13.

He was greatly attracted by a pantheistic form of nature-philosophy, which he embodied in a number of verbose but highly imaginative scientific poems, in which he proclaimed the organic connection between species. In his *Temple of Nature* of 1803 for example, he traces the progress of life from lower to higher forms in what is unmistakably an evolutionary process:

> 'Organic life beneath the shoreless waves
> Was born, and nurs'd in ocean's pearly caves.'

Erasmus Darwin's idea of evolution was intuitive rather than empirically based, but his grandson, Charles, certainly read these poems, and it is difficult to believe that they had no influence on him. Georges Cuvier (1769–1832) was a palaeontologist who amassed far more data than the older Darwin, and built from them theories which were new and intriguing, although often incorrect. Because of the eminence he achieved in French science, and the tenacity with which he asserted his ideas, he became known as the 'dictator of biology'. Cuvier established that the deeper the rock strata, the more unlike living species the fossils became. He realized that whole species of animals had suffered extinction, but supposed that this had been caused by catastrophes such as floods and earthquakes. He was clearly uncomfortable with the idea of successive creations, and proposed instead re-population by migrant species. Like Buffon, he envisaged a fairly short time-span for the age of the earth, in tens of thousands of years. He consistently

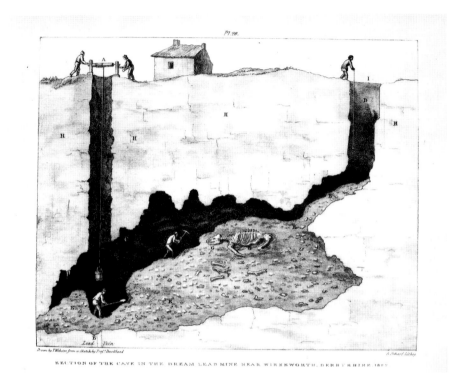

SECTION OF THE CAVE IN THE DREAM LEAD MINE NEAR WIRKSWORTH, DERBYSHIRE 1822

Evidence of the Flood: in 1823 the Oxford geologist William Buckland published an account of bones found in a cave in Derbyshire, which he was convinced were the remains of creatures drowned in the flood; the flood could thus be seen as an explanation of the former existence of animal species which were now extinct. From Buckland's *Reliquiae diluvianae*.
The British Library 443.g.19.

defended the doctrine of the fixity of species, arguing that each animal is perfectly adapted in its own way, and that change would therefore threaten its survival. He felt indeed that the differences between the major zoological groups such as vertebrates and molluscs were so great that they constituted quite distinct life-forms, with no common origin. Cuvier's achievement was to establish the principle of extinction, and to recover and reconstruct many lost species; but he failed to see the true relationship between the living and the dead species. His work provided material for later evolutionists, although he himself was firmly opposed to the idea. The importance of stratigraphy had been recognised since Stensen's work in the seventeenth century, and was developed by Hutton, Cuvier and many others. William Smith (1769–1839) made the first geological map of Britain, and he gave a very explicit statement of stratigraphy in his book *Strata Identified by Organised Fossils*, 1816, in which he showed that certain types of fossil were always within certain rocks, and that therefore the fossil species were contemporary with those rocks. This idea of faunal sequence was later confirmed as a universal principle.

The founding-father of nineteenth-century geography was the Prussian, Alexander von Humboldt (1769–1859), whose work enshrined the principle of the interdependence of nature. Humboldt's famous South American journey of 1799–1804 provided him with material which he digested and published during the following thirty years. Humboldt wished to show that the subdivisions of natural history – zoology, botany, geology, meteorology and so on – all shaped and determined the others, that they formed in fact what we should now call a single science of the environment. The most important result of this study,

Glaciers and the Ice Age.
It was the Swiss Louis
Agassiz who proposed
that the movement of
glaciers had deposited
rocks far from their
source, and he argued
that glaciers had once
covered much of Europe.
From Agassiz's *Études sur
les glaciers,* 1840.
The British Library 1254.l.9.

Humboldt wrote, was 'a knowledge of the chain of connection by which all natural forces are linked together and made mutually dependent on one another, and it is the perception of these relations that exalts our minds'. Humboldt inspired a new era in geography, and his followers emulated him in searching for knowledge in the jungle, the mountain and the wilderness. In his omniscience, his prowess as an explorer, and his sense of the grandeur of nature, Humboldt is the prototype of the *savant,* the scientific hero encountered in the fiction of Jules Verne. Strangely, Humboldt rejected as false one of the grandest geographical discoveries of this era, that of the Ice Ages. The Swiss, Louis Agassiz (1807–1873), studied glacial action in the Alps, and concluded that it was responsible for gravel beds and large, isolated boulders that are found miles from their related rocks. Recognizing that these phenomena were to be found not only in the Alps but in regions far from any glaciers, he suggested in 1837 that an ice sheet had once covered northern Europe and had acted like a vast and complex glacier. Orthodox science could not accept such as strange picture, partly because it could not be reconciled with the cooling-earth theory which had been accepted since Buffon.

By the late 1820s the moment was clearly ripe for a new synthesis in geology, comparable to that performed in chemistry by Lavoisier, and that which Carnot, Joule and Kelvin were initiating in physics. A century of serious geology had produced a number of very important ideas. The word geology was coined in 1779 by another Swiss, Bénédict de Saussure, the pioneer of mountaineering and Alpine studies, and such discoveries as the stratigraphic sequence, the extinction of animals and the igneous or sedimentary origin of rocks were permanent advances. Yet the science was hampered by the assumption that any changes in the face of the earth must have had sudden, catastrophic causes, and above all by an inadequate concept of geological time. It was Sir Charles Lyell (1797–1875) who laid the foundations of modern geology by establishing the two complementary principles that the earth has been formed by the same gradual processes of rock-building and erosion which are active today, and that consequently geological time must be measured in hundreds of millions of years. Lyell was a pupil of William Buckland at Oxford, and began his career accepting the catastrophic geology of his teacher, but in a few short years original observations convinced him otherwise. While still a young man during travels in Europe he observed, for example, rivers in Italy whose deposits had created coastal plains and had silted up ancient ports so that they now lay some miles inland from the sea. In the Auvergne he identified long-dead volcanoes whose lava dated from several different epochs, intermingled with evidence of freshwater deposits from lakes and rivers and many fossil species; this environment, he reasoned, was the record of a process extending not over thousands but over millions of years. In Sicily his study of Mount Etna convinced him that it had arisen not from one great explosion, but had been built up over long ages. These and many other observations suggested to Lyell that the processes that operated in the past were exactly similar to those operating now, and since the time-scale of volcanic life and river deposits could be roughly calculated, activity in such regions must clearly have a history whose length was undreamed-of until now.

By 1828 Lyell had begun work on a book in which these conclusions would be set out with a mass of supporting evidence. In the first part of *Principles of Geology,* published in 1830, Lyell enunciated the principle of uniformity, that the geologist should always try to explain the phemonena of the past by analogy with the present: on this basis, even the greatest changes could be explained given sufficient time. Lyell drew attention to a very significant difference between the fossil sequence of land and sea animals, showing that there was a great continuity in marine life, while land species seemed to have become extinct much more rapidly. He argued that this was because the land environment was in a state of gradual change, that the animals were subject to greater vicissitudes, that new conditions favoured new species, and that in fact the emergence of new species was a permanent

FIG. 4.—*TABLE OF STRATIFIED ROCKS.*

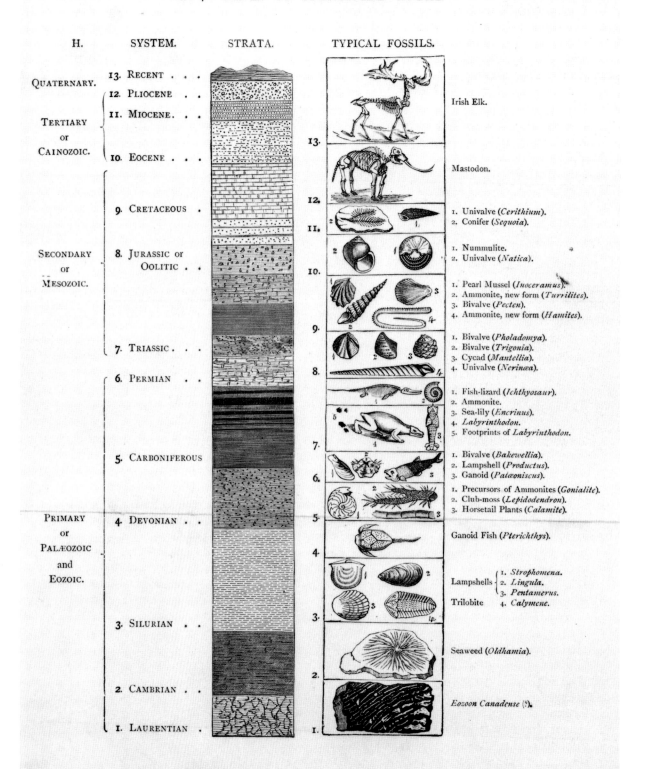

condition of nature. Lyell thus emphasized that the earth and the living world were in a state of dynamic balance, in clear agreement with Humboldt's view of geography. *Principles of Geology* was completed with its second and third parts in 1832 and 1833. Lyell's work created a sensation in its revelation of the age of the earth, but Lyell had avoided any grandiose speculations about the origin of the earth, or about any religious implications his work might have, and the weight of evidence which he had produced was generally convincing. He made the age of the earth a question of evidence rather than one of doctrine. Lyell's achievement was to place geology firmly on a scientific basis by arguing that the natural laws which govern it are the same now as they were in the past, just as those of physics or mathematics are, and that wild hypotheses concerning catastrophic forces

The 'Hydrarchos' on public show, 1846. Palaeontology was one aspect of the scientific revolution which was highly visible and easily appreciated; the extinct monsters of the past have never lost their popular appeal. From *Illustrierte Zeitung,* 1846.

which have now vanished from the earth were irrational and had no place in true geology. Lyell's contemporaries such as Roderick Murchison and Adam Sedgwick helped to fix the geological eras such as Carboniferous and Cretaceous, and the rock series within them, Cambrian, Devonian and so on. To the vast time-scale now attributed to the earth was added the knowledge of the astonishing life-forms it once supported, as palaeontologists began to reconstruct the forms of the fossil animals they were discovering. In England, Dr. Gideon Mantell created something of a sensation with his description of extinct reptiles from the chalk of Sussex, such as the iguanodon. Mantell's work marks the dawn of dinosaur-consciousness, although the word dinosaur itself was coined in 1842 by the naturalist, Richard Owen.

Among those deeply impressed by this new view of the earth was Charles Darwin, who read *The Principles of Geology* during his historic voyage of 1831–36. The story of Darwin's life and his ideas has been told many, many times, in proportion to their profound importance. In contrast to the professional and academic status of scientists in France and Germany, Darwin was still very much the amateur, an English naturalist on the eighteenth-century pattern, living on a private fortune, avoiding publicity or controversy. The voyage of the *Beagle* to South America and the Pacific afforded the opportunity to examine a huge variety of species in their different habitats, while the reading of Lyell revealed to Darwin the concept of nature as a dynamic balance, its processes extending over aeons of time. The diversity of species was the overwhelming problem: why should there

Facing page: Geological strata: the principle that certain types of fossils were always identified with certain rocks, and that therefore those life-forms were contemporary with those rocks, was emerging even before Lyell. By the 1880s the detailed mapping of geological strata and their fossils had become fundamental to geology. From Clodd's *Story of Creation,* 1887.
The British Library 7004.a.2.

be hundreds of species within any family of animals and plants, distinct yet fundamentally similar to each other? There must be an organic relationship between
them, and they must have diverged over time from common ancestors. So much
had been suggested already, as long ago as Buffon and Lamarck, by Darwin's own

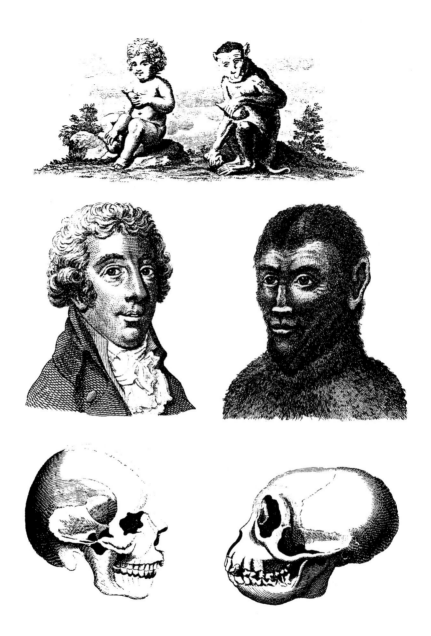

The affinity of man and
the primates: long before
Darwin, similarities such
as these raised questions
about the traditional
doctrine of the fixity of
species. From Sibbly's
Universal System, 1800.
The British Library 1500/871

grandfather, and now by Lyell. The fixity of species defied all reason, yet on
Lamarck's opposite view, the species had no real, stable existence, it was a plastic
condition capable of change in a generation, an idea that neither Darwin nor
Lyell could accept.

   It was after his return to England during 1837–38, while brooding on his notes
and diaries from the *Beagle,* that Darwin discovered a possible basis for the transmutation of species. Darwin himself stated that an important clue was provided
by his reading of Thomas Malthus's essay on human population, with his idea

that a rising population would always press upon the supply of food, resulting in the ever-present threat of starvation. Malthus argued that the fittest – the most resourceful, hard-working and intelligent – would survive in this struggle for existence. Darwin applied this idea to the biological kingdom and extended it over the huge time-scale which he had learned from Lyell, and was then able to conceive of species evolving one into another, always adapting better to their environment, and so surviving while their rivals became extinct. To the mechanism at work Darwin gave the name 'natural selection', by which he meant that slight individual physical differences among animals gave them an advantage in the quest for food; it might be speed or strength or the development of a particular organ or faculty which enabled that individual to survive and flourish, and that very characteristic would be bred with other successful individuals, and reinforced over the generations, ultimately giving rise to a new species. It was not a direct response to the environment in the way that Lamarck had imagined – an animal could not stretch its own neck or beak in order to obtain food – rather the food would be obtained by the animal that was best equipped, who would then survive and bequeath his advantages to his offspring. It was the concept of natural selection, or adaptive evolution, with its essential framework of Lyellian geological time, that explained both the diversity and the similarity of species which had baffled naturalists for so long. Instead of the organism perfectly adapted to its environment which Cuvier had imagined, Darwin's world was dominated by the forces of change and intense competition, competition within species and between species.

Darwin pondered long over his theory, rather as Copernicus had over his in the 1520s and 1530s, indeed there is an interesting parallel between the two theories as revolutions in thought rather than revolutions in evidence. It is not certain when Darwin would have publicized his ideas had another gifted naturalist, Alfred Russell Wallace (1823–1913), not written to Darwin announcing that he had independently reached a similar view concerning the divergence of species by natural selection. Darwin was well aware of the historic importance of these ideas, and, despite his retiring nature, had no wish to see twenty years of original thought pre-empted by another. Papers outlining the theory of evolution by both Darwin and Wallace were read to the Linnaean Society in London in July 1858, and Darwin hastened to compose a full account of them. His book with its graphic, almost self-explanatory title, *The Origin of Species by Means of Natural Selection; or the Preservation of Favoured Races in the Struggle for Life* was published in November 1859. It at once became famous and controversial; among scientists its weight of evidence made it widely accepted, but it was profoundly disturbing to those of orthodox faith. It was clearly quite irreconcilable with the book of Genesis, indeed it appeared to leave no room for divine intervention in nature, which was now seen as a ruthless, unchecked struggle for life, ruled by chance alone, and man as a part of this jungle. Darwin and Lyell both discussed the fossil record of man and placed him explicitly within the system of evolution, indeed they could not do otherwise. But if man was simply an advanced primate, distanced from the apes only by time, what became of man's special relationship to God, his special place in the creation, and what became of the soul? These questions continued to be raised for years after the publication of *The Origin of*

The Machine Age
&❧

Title-page of Darwin's *Origin of Species,* 1859. The sub-title was evocative, enticing and controversial, and the book was a popular bestseller. Along with the texts of Copernicus and Newton, it was one of the handful of books which revolutionized man's understanding of his world.
The British Library C.112.b.12.

ON

THE ORIGIN OF SPECIES

BY MEANS OF NATURAL SELECTION,

OR THE

PRESERVATION OF FAVOURED RACES IN THE STRUGGLE
FOR LIFE.

By CHARLES DARWIN, M.A.,
FELLOW OF THE ROYAL, GEOLOGICAL, LINNÆAN, ETC., SOCIETIES;
AUTHOR OF 'JOURNAL OF RESEARCHES DURING H. M. S. BEAGLE'S VOYAGE
ROUND THE WORLD.'

LONDON:
JOHN MURRAY, ALBEMARLE STREET.
1859.

The right of Translation is reserved.

PUNCH, OR THE LONDON CHARIVARI—May 25, 1861.

THE LION OF THE SEASON.

Alarmed Flunkey. "Mr. G-G-G-O-O-O-RILLA!"

No scientist's work had ever provoked such a public response as Darwin's, and the idea of man as ape was a heaven-sent gift to the humorists. From *Punch* 40, 1861.

The British Library PP.5270

Species; if natural laws could explain the evolution of the earth and of life upon it, what room was left for God? This area of science was plainly more directly threatening to traditional assumptions about man and his world than any discoveries in physics, chemistry or even astronomy had been; indeed Darwin's own wife, Emma Wedgwood, was distressed by his ideas, and never accepted them. It has even been suggested that the persistent and mysterious ill-health which Darwin suffered during the 1840s and 1850s while working out his theory, was of psychosomatic origin, springing from his fear of the consequences of his ideas; later, during the 1870s, he enjoyed the best health of his life.

There were two factors which tended to diffuse the explosive potential of Darwin's work more quickly than might have been expected. The first was that by the 1870s historical criticism was already changing the perception of the biblical narratives, and scholars such as Julius Wellhausen in Germany and Robertson Smith in Britain were interpreting the account of the creation in the book of Genesis as a form of myth, not essentially different from other creation myths of the ancient world; the biblical narratives were now seen as human documents, not as literally the word of God. Among theologians and scholars therefore, who might otherwise have led the attack on Darwinianism, there was already a strong movement towards a more sophisticated view of the Bible and its authority in matters of history and science. The second development was the positive, moral interpretation of evolution associated with thinkers such as Auguste Comte and Herbert Spencer. In this view, evolution offered not a picture of a ruthless jungle, but of progress out of the jungle: life was clearly engaged in an ascent from lower forms to higher, with civilized man the highest form of all. There was indeed order and progress in nature, and although most 'positivists', as they came to be called, were undoubtedly atheists too, this view of evolution could easily be transformed from a secular into a religious belief. This process of moralising the theory of evolution is very reminiscent of the way in which Newtonian science had become the basis of a rational theology in the eighteenth century. The enormous potential of the theory of evolution, philosophical and social as well as scientific, explains the central position which it has occupied in western intellectual life ever since Darwin.

There was however one acute scientific problem raised by Lyell's geology and Darwin's zoology which was not readily solved, and it came from the world of physics. The general theory of the origin of the earth was the nebular hypothesis, in which the earth solidified into an intensely hot molten mass, which had been gradually cooling ever since. The leading English physicist, Lord Kelvin, had estimated the duration of this cooling process, and on that basis he stated the age of the earth to be at the most 25 million years, which, great though it was, was insufficient to allow for the processes described by Lyell. This controversy between geology and physics would not be resolved until the early twentieth century, when radioactivity as a source of heat within the earth was discovered, and Kelvin's view of the age of the earth became invalid. This problem was an example of the maturity of science, where established laws in one body of knowledge might be applied to criticize or dispute those of another; out of the reconciliation of the separate disciplines grew the important principle of 'consensual' knowledge.

As the nineteenth century advanced, the life sciences became a focus of research in more and more laboratories and hospitals throughout Europe, and soon in America too. In particular the German university system and the state-funded science of France encouraged a new form of pure research, in which the rigorous aims and models of the physical sciences were applied to the living world and the human body. The age of intense specialization had not yet arrived, so that discoveries in chemistry, botany and zoology might all contribute new perspectives on each other and on human medicine. As in the physical sciences, the intellectual current was predominantly away from theories and hypothetical entities, towards physical and clinical experiment. Anatomy and structure had been thoroughly investigated over many years, but the systems and functions of the living organism – the science of physiology – was still waiting to be unveiled. The institutionalization of research in biology and medicine in scores of centres from Scotland to Hungary, from Russia to Spain, created an array of schools and of outstanding teachers, passing on their own traditions and methods, so that biological science became more and more a professional, cumulative enterprise. This widespread programme of deliberate laboratory research was something new in medicine, and it culminated in a new understanding of the life process and of disease. Yet amid all this academic professionalism, the life sciences were still fluid enough for one of the most original and far-reaching discoveries to be achieved by an amateur botanist: Gregor Mendel laid the foundations of the science of genetics while working in private and virtually isolated from the scientific community.

The new era in physiological science was proclaimed in Paris by Francois Magendie (1783–1855), who felt that medicine had lagged behind the physical sciences because it had been lured into speculation and hypothesis, instead of analysing basic activities like nutrition or the sensory-motor system. Adopting techniques learned from chemistry, Magendie began experiments to isolate the active ingredients in herbal and animal drugs, such as strychnine, and thus may be seen as one of a founders of modern pharmacology. Magendie was involved in a priority dispute with the Scots physician, Charles Bell, over the discovery that the motor and sensory functions are distinct in the medulla, divided respectively between the anterior and posterior parts of the stem; this is now known as the Bell-Magendie law. Magendie's counterpart in Germany was Johannes Müller of Berlin (1801–1858) who influenced a whole generation of scientists. While still a young student, Müller demonstrated the respiration of the mammalian foetus via the umbilical cord. It was Müller who first drew attention to a fundamental puzzle about the nervous system, that each sensory system acts in a specific and limited way, although the nerves are physically indistinguishable; nerves from the eye and the ear respond only to light and sound, and so on, a phenomenon which Müller termed the 'law of specific nerve energies'. Both Magendie and Müller could advance physiology only through the continual vivisection of animals. On a trip to England in 1824, Magendie demonstrated the effects of severing various nerves in living dogs, and these events marked the beginning of a campaign against vivisection in England, which found however no echo in France. Magendie's approach was continued by his pupil, the man usually regarded as the founder of modern clinical physiology, Claude Bernard (1813–1872). Bernard researched principally in the nervous and digestive systems, discovering for example the function of the liver in building up sugar and releasing it into the blood as glycogen. Bernard showed that the body was regulated by many such complex chemical processes, and this led him to his influential concept of *le milieu intérieur,* the 'internal balance' of the body which is essential to life. The practice of medicine held no appeal for Bernard, and he was entirely a man of the laboratory, austere and friendless, who became estranged from his wife and daughters, partly it is said because of their disgust at his animal experiments. Bernard brought to fruition Magendie's aim of freeing medicine from any suspicion of vitalistism, and replacing it with a strict determinism, based on physiological analysis.

A centuries-old dispute was finally settled and a new phase of biological history was opened when Karl von Baer of Königsberg (1792–1876) succeeded in 1826 in identifying a mammalian ovum. It was plainly a single, undifferentiated cellular unit, and thus the myth of preformation was laid to rest, and the principle of epigenesis was established. Baer wrote that 'every animal which springs from the coition of male and female is developed from an ovum, and none from a simple formative liquid'. This discovery was a starting-point for a new, systematic study of animal development, one of the outstanding focuses being the problem of the cell. Even in the early years of the century the German naturalist Lorenz Oken (1779–1851) had conjectured the existence of *Urschleim,* a fundamental fluid which he thought was contained in minute spheres within living tissue; outside Germany *Urschleim* became 'protoplasm'. In 1821 the Italian microscopist Giovan Battista Amici (1786–1868) observed the fertilization of pollen in the pistil of a flower, while in 1831 the Scottish botanist Robert Brown (1773–1858) identified nuclei inside the microscopic spheres which he now referred to as cells.

All these were important clues, and attention came to be focussed on the cell as the possible elementary unit of life because it was clearly common to both plants and animals. François-Vincent Raspail (1794–1878) was a self-taught public health activist, commemorated in innumerable street-names throughout France; a thorn in the side of the medical establishment, he was often imprisoned for his political activities. Raspail spoke of the cell as 'a type of laboratory of cellular tissues that organize themselves and develop within its innermost substance'. But it was two of Johannes Müller's pupils, Jacob Schleiden (1804–1881) and Theodor Schwann (1810–1882), who systematically described the manner in which different types of tissue are composed of distinct cells which yet share common characteristics. Some cells are independent and isolated, as in the blood; some are pressed together with well-developed walls, such as bones and teeth; while others are elongated into fibres such as tendons and ligaments. Beneath this diversity, Schleiden and Schwann became convinced that the cells' life, although to some extent independent, was subjugated to that of the organism as a whole, and that they were indeed the fundamental building-blocks of life. This insight was extended by Rudolf Virchow (1821–1902) who identified cell division as the mechanism of organic growth, and enunciated the doctrine that every cell arises from another cell. Virchow understood that chemical processes were at work within the cell, and that disease was reducible to cell pathology, malfunctions of the basic units of life. Virchow was a political radical very concerned with public health, who once prescribed the remedy for a typhus epidemic as 'democracy, freedom, education and prosperity', and it is tempting to trace a political inspiration for his theory of the organism as a federation of individual cells, ideally thriving in peaceful cooperation, but always threatened by the malfunctioning of their fellows. Science became for Virchow a quasi-religion, with medicine seen as the science of man, both physical and social, but he became extremely pessimistic about the direction in which German society was moving. In later life he became increasingly interested in history and anthropology, and accompanied the great archaeologist Schliemann during some of his excavations of Troy.

The mature cell theory opened the way for biochemistry and genetics, and interesting parallels could be drawn with the way that atomic theory, by reducing matter to its constituent elements, permitted the rapid growth of chemistry and physics. The process of cellular discovery that began perhaps with von Baer, reached an important milestone between the years 1877 and 1879, when two German scientists working independently, Hermann Fol (1845–1892) and Wilhelm Hertwig (1849–1922) observed the penetration of the sperm into the egg. Both these men had found that sea-urchins were ideal subjects because of their transparency, and with the aid of injected stains, they were able to see that within minutes the nuclei of the two cells fused to form one fertilized egg. Thus a controversy which had endured since Aristotle concerning the process of generation, the primacy of the male or female principle, was at last settled in what, in retrospect, seems the simplest possible way, that two seeds became one; it seems obvious and elementary, but this knowledge is now little more than a century old. In the 1880s, Edouard van Beneden (1846–1910) of Liège observed the next crucial stage in growth, the division of the nucleus. He drew attention to the elongated particles, later termed chromosomes, which the microscope now revealed, and which were replicated when the nucleus divided.

The growth of bio-medical science was rapid, but unifying principles were still relatively few, so that the possibilities of error were enormous. One of the most enigmatic figures in nineteenth century biology is Ernst Haeckel of Jena

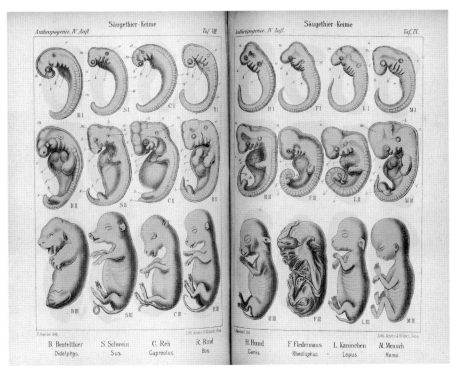

Haeckel's embryology: Ernst Haeckel conceived the idea of embryonic recapitulation, that the human embryo repeats in a few months the evolution of life from primitive forms. The drawings with which he supported this view are now regarded as having been designed to fit his thesis rather than to reflect reality. From Haeckel's *Anthropogenie* 1891.

The British Library 7006.h.9.

(1834–1919), perhaps the most renowned expert in his field, showered with academic and popular honours. Haeckel was a pupil of Müller, and in his turn he trained many leading biologists, and wrote highly influential works of popular science. Haeckel took as his special field evolutionary morphology, and to him is

Portrait of Louis Pasteur
in his laboratory.
Pasteur's conquest of
the invisible sources of
disease gave him a fame
unprecedented among
scientists. From *The
Graphic*, 52, 1895.
The British Library,
Newspaper Library, Colindale

due the theory of embryonic recapitulation, the idea that the human embryo re-
enacts in a few months the evolution of life from its beginnings as a single cell.
Haeckel published pictures of the embryonic development of many species which
appeared to support this theory, but which are now regarded as inaccurate.
Haeckel was to some extent a very late disciple of German Nature-Philosophy,
seeking an all-embracing picture of the evolution of man which went beyond the
available evidence, indeed he himself advocated a 'philosophical empiricism' as
the ideal method of science. The extent to which system-building and speculation
led Haeckel to misinterpret his evidence, or even to falsify it, is still debated.
More important if less grandiose was the system-building of the founder of
organic chemistry, Justus von Liebig (1802–1873), who analysed the carbon
cycle from air to plant to animal, a classic example of the conservation of matter,
showing the different realms of nature to be linked in a process of synthesis and
dissolution of a small number of elements.

The focal, heroic figure of nineteenth-century biomedicine was undoubtedly
Louis Pasteur (1822–1895), who demonstrated the existence of an unsuspected
form of life – the microbe – and formed a new model of disease as the invasion of
the body by these external agencies. Infection and contagion had of course long
been understood on a practical level – that is why lepers and plague victims were

segregated in the ancient and medieval world – but it was always assumed that it was the manifestations, the effects of disease which were being avoided, while the internal cause remained a divine or perhaps an astrological mystery. A true understanding of the agency of disease ranks as one of the great intellectual problems of science, and it was Pasteur who solved it. Pasteur's method was always intensely practical, experimental and tenacious, yet his work was also strongly directed by intuition, and there was even a hint of vitalism in his thought, so that when he arrived at a conclusion he defended it fiercely, even blindly. The exact process that led Pasteur to his historic germ theory was extremely complicated, arising from a long study in the 1850s of the nature of fermentation. The orthodox view was that fermentation was a purely chemical process in which sugar was turned into alcohol, acids and gases. The fact that it occurred in the absence of air, when oxygen was known to be essential to respiration and combustion of any kind, supported this view. Pasteur found however that it was impossible to embody this process in any chemical equation, for its by-products were many and complex, and the nature of yeast resisted chemical analysis. This lay behind his key insight that yeast was a living organism which survived without air by extracting oxygen from the sugar, and grew by assimilating carbon and nitrogen from the same source. This was conclusively proved when yeast is exposed to air, for it grows but does not ferment sugar; only in the absence of free oxygen does fermentation take place. This concept of 'life without air' was new and highly controversial in the scientific world, but was central to Pasteur's discovery of this unsuspected realm of the microbe. It offered the key to 'the secret and mysterious character of all true fermentations, and possibly that of many normal and abnormal actions in the organization of living things'. Among these actions was putrefaction, which Pasteur now brought within the scope of micro-organisms; the decomposition of animal and vegetable matter was effected by their recycling in their own life-functions of carbon, hydrogen and nitrogen. 'Life takes part in the work of death in all its phases', wrote Pasteur, indeed death can be seen as essential part of the universal cycle of the chemical elements; the characteristic odour of death was merely the release of sulphur in the form of gases. Pasteur recognized that there are two basic type of microbes, the anaerobes which function without air to produce fermentation, and the aerobes which live at the surface and cause oxidation, producing water, ammonia and other products.

Death enthroned in his kingdom, 1835: the satirist places the work of the doctors themselves among the causes of death before the era of asepsis and anaethesia.

Where were these organisms normally to be found, and how did they spread? Pasteur conducted a series of rigorous experiments with organic matter in air, in sealed containers, in containers that had been boiled, and in filter-controlled containers, all of which demonstrated beyond doubt that earth, air and water were filled with an immense and invisible population of living microbes, many of which were detectable with the microscope, and which were responsible for basic chemical processes in the animal and plant kingdoms. Pasteur provided the final answer to the ancient problem of spontaneous generation, which, despite being laid to rest by Spallanzani in the seventeenth century, had more recently been revived, curiously because of the rise of scientific chemistry. The discovery that oxygen was essential to respiration had prompted biologists to look again at Spallanzani's experiments, and to suggest that it was simply the lack of oxygen which prevented spontaneous generation from taking place. In 1859 a respected naturalist, Félix Pouchet, published a long and detailed defence of spontaneous generation, which was now of more than scientific interest, for the issue had become bound up with the contemporary argument between the materialist and religious view of life. Materialists were attracted by the idea of spontaneous generation as providing evidence that life was an impersonal, physical process which could begin at any time under the right conditions, while the traditional religious view was that the creation of life was a unique and divine event. Pasteur was a man of strong religious faith, and he fiercely attacked spontaneous generation as threatening the very concept of God's creative power. The French Academy of Sciences entered the debate in 1860, offering a prize for a conclusive answer to the problem, and Pouchet and Pasteur mustered their forces for the gladiatorial struggle; this was science contending for the soul of nineteenth-century man in the most public way. Pasteur proved first that air will produce putrefaction or fermentation in previously sterile organic matter; second however, that it is not the air alone, but something contained in the air, something which can be carefully filtered to prevent this activity; third that this activity is reducible under the microscope to tiny bodies, sometimes motile; and fourth that environmental factors such as temperature directly affect the density of microbes in the air – he experimented in the Jura mountains up to heights of 2000 metres. Most of these results were embodied in his historic paper of 1861, 'Memoir on the organized corpuscles which exist in the atmosphere – a critique of the doctrine of spontaneous generation'. Pasteur won the Academy's prize, and the unfortunate Pouchet was vanquished.

Pasteur's fame was now considerable, and he was drawn into public service by a series of government invitations to solve technical problems in the wine industry, the silk industry and in livestock and agriculture. The story has been told many times how he saved the French silk industry by isolating the germs of two diseases which were destroying the silkworms; how he showed that heating wine or milk to the correct level would destroy the bacterial agents which caused fermentation and souring; how he was able to combat the animal diseases chicken-cholera and anthrax, the latter by a highly public demonstration in 1881 of the power of the vaccine he had developed. In this work he learned directly from Jenner's earlier achievement with smallpox vaccine and the concept of immunity, but Pasteur developed the important technique of attenuating the infected material until the correct level for a course of immunization was reached. This was experimental work, in itself dangerous to life, acceptable perhaps in the animal field, but Pasteur hesitated long before entering the field of human medicine and testing the full implications of his discoveries. Many in the medical establishment rejected the possible application of germ theory to human disease; the notion that a large powerful creature could be attacked and killed by agents too small to be seen was ridiculed, and the presence of these microbes was explained as a result of disease, not its cause. By the early 1880s however, Pasteur, and others such as Robert Koch in Berlin, had concluded that separate pathogenic agents were the cause of many if not all diseases, whether anthrax or smallpox, puerperal fever or tuberculosis, and were arguing that this new model of disease should lie at the very centre of medical science.

Perhaps no event could be calculated to display the triumph of the new theory more dramatically than the conquest of rabies, a disease that was mysterious, horrifying and invariably fatal. Pasteur had witnessed it, and the excruciating treatment of cauterizing the wound, during his childhood in the Jura. He experimented with the disease for five years from 1880 onwards, and had indeed extracted a vaccine from infected dogs which gave a measure of protection, although its effects were still uncertain. In July 1885 a nine-year old boy, Jacob Meister from Alsace, was brought to Pasteur having been severely bitten by a rabid dog, and considered doomed by the doctors. Members of the government rabies commission urged Pasteur to use his vaccine, and with great foreboding he

ESSENCE OF PARLIAMENT.
EXTRACTED FROM
THE DIARY OF TOBY, M.P.

HOUSE of Commons, Thursday (anticipatory). — Members all back as delighted as if they were going away. Everybody shaking hands with everybody else. PETER RYLANDS doing the honours of the place, as it were ; quite in boisterous spirits.

"Another good Under-Secretaryship gone wrong," DRUMMOND-WOLFF slily whispers in his ear. "You'd better come over and join us."

"Thanks ; but I'll wait a bit longer," PETER says. "CHILDERS was all very well at the War Office ; it's different at the Treasury. I give him six months there, then there may be a call for a man who has finance at his finger's ends, is trusted by the country, and is a pretty fair speaker."

BRADLAUGH in high spirits. Tells me he's been round spending half an hour with GOSSET practising the steps. Sergeant-at-Arms, it seems, who has not forgotten his old skill, wants to reverse when they waltz backward from the Mace. After the practice of three Sessions, BRADLAUGH can do the forward step well enough, but finds it hard to reverse. Still means to try.

"The eyes of the country are upon us," he says, "and we must do the thing well."

Black Rod arrived shortly after two o'clock. Door shut in his face as he

Politicians as microbes – bizarre life-forms distinct from the world of humans. Pasteur's revelation of a hidden world of teeming organisms in the earth and the air was an intellectual revolution in bio-medicine. The microbe, like the dinosaur and the atom and the subconscious mind, would become scientific facts which also entered the popular imagination. From *Punch* 84, 1883.

The British Library PP5270.

consented. After ten days the boy was receiving injections of the most virulent rabies vaccine, but the disease did not appear, and four months after he had been bitten, his life was assured. Other cases followed, and despite lingering doubts among doctors and some isolated failures with vaccines, the tide of Pasteur's fame swelled to new heights; the Pasteur Institute was founded in Paris, while streets, schools and even villages were named in his honour, and the Pasteurian model of disease was triumphant. His work had a mythic quality of challenging and destroying a dark and hitherto invincible cause of suffering and death. The social significance of Pasteur's work has to be seen as an aspect of the religion of progress: it was widely believed that a new era in history had opened, in which science would defeat one by one the enemies of humanity. In one of his last public statements, Pasteur spoke of his 'invincible belief that science and peace will triumph over ignorance and war'. Had he lived a few years longer, he would have seen the ruin of this creed, but Pasteur is worthy of his fame, and he belongs with the greatest scientists – Copernicus, Newton and Darwin – who have changed forever our understanding of the physical world or of man's place in it.

Robert Koch of Berlin (1843–1910) was some twenty years Pasteur's junior, but began his own research into the microbiology of disease independently. In 1882 he succeeded in isolating the bacterium which caused that scourge of nineteenth-century Europe, tuberculosis, but his announcement that he had developed a vaccine against it was premature. Yet even this highly-publicized disaster could not turn back the germ theory of disease, and Koch was among those who spearheaded research into the major killers which Europeans encountered in their overseas colonies as well as at home, such as cholera, malaria, dysentery and typhus. The simplest and most dramatic application of Pasteur's teaching was seen as early as the 1860s when the Glasgow surgeon Joseph Lister (1827–1912) was immediately convinced by the germ theory, and began using carbolic acid as a disinfectant. In Vienna Ignaz Semmelweis (1818–1865) actually anticipated Pasteur and Lister by inaugurating an antiseptic regime, especially in maternity wards, where mortality from puerperal fever was catastrophically high. Semmelweis had noticed that doctors proceeded from autopsies to deliveries, and he drew the correct conclusions; but this was in the 1850s before Pasteur had articulated his theory, and the resistance from the medical profession was so great that Semmelweis retired a broken man. Asepsis was one of the twin pillars of nineteenth-century medical practice, the other being anaesthetics, and together these revolutionized the possibilites of surgery. If the primitive use of alcohol is discounted, anaesthetics were pioneered in the 1820s with nitrous oxide (laughing gas) and then ether, before Sir James Simpson (1811–1870) in Edinburgh settled in 1847 on chloroform. The technical history of medicine in the years 1840–1900 is the story of the great teaching hospitals in London, Paris, Edinburgh, Berlin and elsewhere, lorded over by their fierce, omnipotent professors, who were in reality still exploring half-understood realms of pathology and surgery, and whose names are commemorated in the diseases they described – Addison's, Bright's, Parkinson's or Paget's. The heroic but improvisatory state of the profession in these years is beautifully evoked in the medical fiction of Arthur Conan Doyle, himself an Edinburgh medical graduate, with stories of first operations, crude electro-therapy, the horrified discovery of inherited syphilis, the entry of women into the profession, miraculous cures discovered by chance, and daring surgeons whose knives 'cut away death but grazed the springs of life in doing it'. In this culture, medicine, as well as being a nascent science, was perhaps another realm of discovery, calling for courage and character rather than analytical genius.

On the more scientific level, the foundations of organic chemistry and biochemistry were laid before the end of the century. Paul Ehrlich (1854–1915) showed that the cell operates through normal chemical processes, and that antibodies disable identifiable toxins in the bloodstream in the same way. In this pre-antibiotic age, he pioneered chemical therapy as an alternative to the vaccine

therapy of Pasteur, and discovered salvarsan, a derivative of arsenic which
destroys the agent of syphilis. Emil Fischer (1852–1919) investigated the part
played by in human physiology by carbohydrates and proteins, chemically synthe-
sizing the former and analysing the latter. In 1878 Willhelm Kühne coined the
term 'enzyme' for the numerous biochemical catalysts which were being discovered, and Fischer proposed the correct lock-and-key model by which they operate
– that they consist of asymmetric molecules which bind specific chemicals
together but are themselves unaffected – thus explaining why a tiny quantity of an
enzyme can participate in repeated reactions. All this was highly specialized
research in professional and academic centres, and its practitioners became stars
honoured in the new intellectual Olympics – the Nobel Prize list. Yet one of the
most significant developments in biological research had taken place in private far
from any laboratory or hospital, and it had gone unnoticed by the scientific
world.

Gregor Mendel (1822–1884) spent some ten years from 1855 onwards in a
series of experiments in the garden of his Augustinian monastery at Brno, trying
to determine the pattern of heredity. Mendel had studied science, including
mathematics and biology, at the University of Vienna, but did not take a degree,
and he was more of a naturalist on the Darwinian pattern than a professional scientist like Virchow or Pasteur. It is uncertain if Mendel had a pre-conceived
theory which he wished to prove, but he certainly planned his experiments
methodically, raising and examining almost 30,000 pea plants which he carefully
pollinated himself, and recording the inheritance pattern of seven distinct characteristics, such as height, colour, leaf shape, and so on. The essence of what
Mendel discovered is this: that when pollen from, for example, yellow peas was
used to fertilize green peas, all the offspring was yellow. This was already interesting, for common sense would expect the colours to mingle, but a greater surprise
followed, for when the second-generation yellow peas were self-fertilized, green
re-appeared in a small number of plants; moreover Mendel established that it re-
appeared in a rather precise ratio of one in four. This pattern was repeated with
the other characteristics which Mendel observed – height, leaf shape and so on.
Mendel's genius lay not only in experimentation, but also in deduction, for he
drew some unexpected but brilliant conclusions. The most important were that
each characteristic is determined by a *factor* which exists independently of the others; that these factors come in pairs, one from each parent; that some of these
factors dominate the pair during one generation, but that the recessive factors are
not destroyed, and may re-appear if combined in a suitable ratio; and, the most
general and important idea of all, that these factors preserve their character intact
as they are transmitted through the generations – they are not mingled or transmuted. By statistically analysing a large volume of data, Mendel succeeded in
formulating laws where before only random phenomena had been perceived.

Mendel could not know whether these 'factors' had some actual physical existence, although he must surely have suspected it; he did not enlarge his findings
into a theory or system, but was content to report his results. Had these been
received with enthusiasm, perhaps he would have extended his work, but they
were not. He announced his results to the Natural Sciences Society in Brno in
1865, and they were published in the Society's journal in 1866, but he received
no response from the scientific community. Discouraged, Mendel published
nothing more on biology, although he did continue his researches into heredity,
and his work lay forgotten for a generation. When it was rediscovered, its validity
was recognized, and it formed the basis of modern genetic theory, although need-
less to say human genetics is more complex than the seven characteristics of the

peas studied by Mendel. The relationship between Mendelian genetics and Darwinian evolution was problematic however; on the one hand Mendel's 'factors' (genes) were seen as providing a mechanism for the inheritance of characteristics, but on the other hand it became more difficult to account for transmutations when it was known that these characteristics were transmitted intact, without blending or modification. This problem was accentuated by the influential theories of August Weismann (1834–1914), who emphasised the separation of the reproductive cells from the rest of the body. What Weismann called the 'germ-plasm' is transmitted from generation to generation, unaffected by its biological environment. Weismann was extremely prescient in his belief that this germ-plasm was located in the nuclei of the reproductive cells, and his theory has been seen as a prophecy of DNA, but it contained echoes of the old idea of preformation, and Weismann was much exercised to explain how variation within species and between species could originate. The harmonization of evolution and genetics has been one of the challenges of modern biology.

THE NINETEENTH-CENTURY ACHIEVEMENT

Until the early nineteenth century it would be possible to write secular history – the history of government, war, social and personal life – without reference to science. Even major innovations such as ocean navigation or the invention of printing, while they were great practical feats, had no basis in theoretical science, and no inevitable extension into new technologies. By the end of the nineteenth century however, science had transformed the way people lived and thought to such an extent that it had become the shaping force of western culture. If one had to characterize the achievement of nineteenth-century science, it could be termed the discovery of the invisible. By experiment and by reasoning from the particular to the general, the revolutions in physics, chemistry and bio-medicine had revealed hidden patterns and forces in nature whose very existence had been unsuspected by the leading minds of the eighteenth century. Yet unlike the work of Copernicus or Galileo or Newton, this did not remain solely an intellectual revolution: the new element in the nineteenth century was the way that science became visible through technology, through the chemical and electrical industries, and through the conquest of hitherto fatal diseases. The consequences were enormous and they they reached far beyond the laboratory; to give just one example, the rise to world power of Germany, Japan and the United States, with all the consequences which flowed from it, would have been inconceivable without the technical leap-forward which gave those countries new industries, new weapons, and new political ambitions.

In man's long search for certainty of knowledge, there was a strong feeling that science was now able to provide it. The most innovative ideas of the nineteenth century were scientific ideas – evolution, germ theory, the conservation of energy – and its great artefact was the machine. For the first time outside France, where overt atheism had developed in the eighteenth century, science became conscious of its intellectual power to challenge or replace religion in explaining the world. As one Victorian scientist wrote, 'We claim, and we shall wrest from theology, the entire domain of cosmological theory'; and again, 'Evolution is the manifestation of a Power absolutely inscrutable to the intellect of man'. The author of these remarks was John Tyndall, founder of the journal *Nature,* who was more of a pantheist than an atheist, seeing science as a revelatory power, almost a new faith. The very word 'scientist' had been coined in 1837 by William Whewell, a

Cambridge don, and author of the first history of inductive science, which now at last replaced the venerable term 'natural philosophy'. In England, experimental science belatedly entered the university syllabus in the new university of London in the 1820s, and, after much heart-searching about the dangers of secular education, even Oxford and Cambridge capitulated in the 1870s. The scientist as a character entered the popular mind, Doyle's Professor Challenger or Verne's Professor Lidenbrock: omniscient, egotistical, irascible, of iron will and mathematical habits. Yet far from being automata, the greatest scientists, Darwin or Pasteur for example, had a strong intuitive sense to lead them to the correct explanation, however fantastic it might appear, a sense arising both from their belief in nature's limitless powers, and their belief in the capacity of the human mind to penetrate them. The models of the conservation of energy and the conservation of matter had placed two instruments in man's hands which enabled him, theoretically, to account for all conceivable change in the natural world, for change was reducible to the spatial re-arrangement of its elements. This was the model envisaged by Helmholtz, and it lies behind Rutherford's phrase that all science is physics or stamp-collecting: it is the analysis of atoms and the forces that control them, or it is mere description.

The importance of science in this era is most clearly evident in the rise of the science of society, the idea that collective human experience is subject to subtle but all-pervasive laws, just as physics or biology are. This idea pre-dates Darwinianism, but its first appearance in the thought of Hegel between 1800 and 1820 was certainly indebted to the general concept of evolution, which was already current. Hegel's central argument was that change in both the natural and human worlds is governed by a metaphysical process in which higher forms and entities are continually emerging from conflicts within lower forms. This process could be seen at work in nature, in society and in the intellect. Hegel was much more interested in theology than science, but the power and flexibility of his model was such that it was capable of appealing to Christians and atheists, capitalist and Marxists, scientists and mystics, all of whom might claim that their special system *was* the higher form. More directly indebted to the rise of science was the 'positivist religion' of Auguste Comte, who saw human thought and history as progressing through three distinct phases, the theological, the metaphysical and then the highest, scientific phase. This vision of human history was provided with a biological mechanism by the mature theory of evolution, while the most complete application of the concept of process to human society was in the thought of Karl Marx, who was convinced that his analysis of history was no mere theory, but was the expression of scientific laws. 'History itself', wrote Marx, 'is an actual part of natural history, of nature's development into man: natural science will in time include the science of man.' In the age of the machine, the greatest machine of all was human society, and the aim of bringing society under scientific control was another of the great ideas of the nineteenth century, and it was possible for even the most conservative to argue that science rather than social revolution would bring health and prosperity.

The weakness of this whole form of social Darwinianism was its subjectivity: socialist atheism is certainly a different form of thought from Christianity, but who is to judge whether it, or any other creed or system, is 'higher'? More serious still is its inadequate concept of explanation, its faith that when something has been expressed in scientific terms, it has been explained. In fact scientific laws – Boyle's law of gases, Dalton's atomic theory, Joule's law of electrical heat, Mendel's laws of inheritance – are powerful tools enabling us to predict effects; they restate in systematic form what might otherwise be seen as random events;

but they do not, ultimately, *explain* anything. True explanation involves purpose – the question why things happen as they do, and not otherwise. But the study of purpose in nature was not part of science, whose aim was the adequate description of things and processes; it was the concern with purpose that lay behind religion and behind myth. In reality, scientific explanation amounts only to ever-more-detailed description, to description in new terms satisfying to each age. Classical theories are models which are valid under most conditions, but cannot be seen as representing final truths. Science is thus a *provisional* report, and however convincing it may appear to contemporaries, it is capable at any moment of being revolutionized, as other provisional theories – the geocentric universe or the four elements, phlogiston or ether – clearly show. The nineteenth-century faith in progress would not survive the early decades of the twentieth century, and discoveries in science in those years would undermine the apparent certainties of matter, space and time which science had arrived at, and make the search for purpose and explanation still more urgent and elusive.

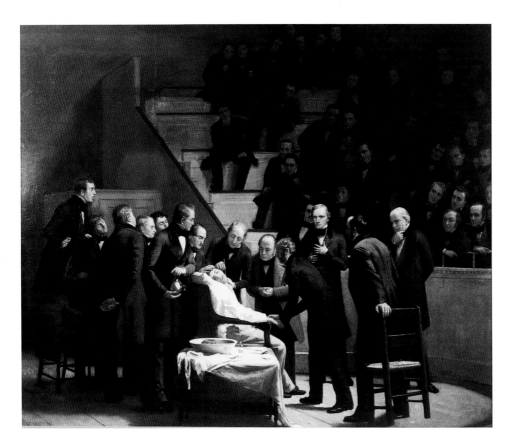

The first public
demonstration of
anaesthesia during
surgery, at the
Massachusetts General
Hospital in 1846.
From a painting by
Robert Hinckley, 1887.
Boston Medical Library,
Countway Library of Medicine

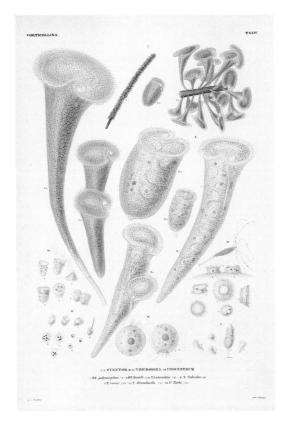

Protozoa: the compound
microscope revealed
unsuspected forms of life,
even before the discovery
of the microbe.
Ehrenberg's book
Infusiontierchen of 1838
was one of the most
impressive atlases of
protozoa ever published.
The British Library 789.i.ll.

The galaxy M74 in the constellation Pisces.
Science Photo Library

Chapter Eight

TWENTIETH-CENTURY SCIENCE: THE NEW LABYRINTH

&❧

'The ancient covenant is in pieces. Man knows
at last that he is alone in the universe's unfeeling
immensity, out of which he emerged only by chance.
His destiny is nowhere spelled out, nor is his duty.
The kingdom above, or the darkness below: it is
time for him to choose.'

Jacques Monod

In the few years between 1900 and 1930, a succession of physicists put forward theories which completely overturned all previous conceptions of the nature of matter, the forces which organise it, and the structure of the universe. These theories reached beyond the laboratory, for they challenged established philosophical ideas about space, time and causation, and about the very possibility of objective knowledge. They would also become the starting-point for new technologies which would deeply affect world politics and the conduct of social life. These theories – the structure of the atom, relativity, and quantum mechanics – created a form of science which, for the first time, inhabited an intellectual world of its own, in a way that the science of Galileo or Newton, Helmholtz or Kelvin had not: they seemed to dissolve the normal human perception of reality, and open a gulf between the truth of everyday life, and the truth of science. They did this through their revelations of the infinitesimal, inconceivable scale of matter's ultimate particles, the unpredictability of the forces that govern them, and the dissolution of a fixed framework of space and time. Where classical physics from Newton onwards had apparently rationalised and unified nature's forces, and had thus opened new bridges to philosophical and even to religious thought, early twentieth-century physics erected structures that were unique to itself, and apparently *irrational*. If nineteenth-century physics was the discovery of the invisible, the physics of the twentieth century has seemed to be at times the discovery of the impossible. It has led us into a new labyrinth, where every turning leads merely to another, and every answer throws up a fresh mystery. This new physics was largely the creation of German scientists, with important contributions from England and France. Was it an accident that this intellectual revolution took place during the years when so much in European civilization was also dissolving? The social and political order that had flourished in the nineteenth century was all but destroyed in the Great War, as was the religion of progress, while literature, music and art were overwhelmed by challenges to formal structure, and by the desire to reveal hidden and possibly destructive forces.

The age was one of excitement and uncertainty, and science played a central role in the Faustian drama of Europe between 1900 and 1940.

The new physics created in this era is the physics with which we now live – the ideas are contemporary and familiar, and they are now taught to each new generation of scientists. What follows will therefore be not so much an exposition of those ideas, as a brief chronicle of the people and circumstances involved in producing them. The ideas themselves are often strange and paradoxical, and their proofs rest on advanced mathematics. It has to be recognized that there now exists a gulf of language between the scientific community and the world at large, which the historian of ideas simply cannot bridge, because history has merged into the present.

The structure of the atom and the nature of atomic processes came to be understood, at least provisionally, by around 1910, and that understanding was reached through the merging of several pathways of research and accidental discovery. The physical reality of atoms was by no means universally accepted, even in the era of Mendeleyev, and the atom was still regarded by many scientists as a working hypothesis. Yet the first attempt to estimate the size of atoms was remarkably accurate. In 1865 Johann Loschmidt of Vienna made some very sophisticated calculations using gas kinetics to arrive at the figure of 10^{-7}mm (0.00000009 mm) as the diameter of a molecule of air. Much later, in 1908, the French chemist, Jean Perrin, devised a delicate experiment to show that the response of tiny particles in water both to gravity and to thermal currents was directly proportional to the atomic weight of their substances; after Perrin, few scientists continued to doubt the physical reality of atoms, but it was now clear why even the most powerful microscopes could not hope to reveal individual atoms: they were far smaller than the wavelength of light itself.

One of the most important pieces of apparatus used by nineteenth-century physicists was the electrified vacuum tube, or the tube containing rarified gases, in which both the spectra of the gases and the electrical discharges could be studied; these techniques led to many unexpected results. In Würzburg in 1895, Wilhelm Röntgen (1845–1923) discovered that such a tube was emitting unknown rays which could not be reflected or refracted and which, strangest of all, seemed to penetrate solid matter. They were related in some way to light, for they reacted with light-sensitive chemicals, so that Röntgen was able to use them to make photographs of objects inside closed boxes, and most spectacular of all, the bones of the human body. Röntgen was unable to explain what they were, but these 'X-rays' caused a scientific and public sensation, and researchers in laboratories throughout Europe began to study them. In Paris in 1896, Henri Becquerel and Marie and Pierre Curie found that similar rays were produced continuously by uranium and by radium, and coined the termed 'radioactivity' to describe their emission. Becquerel and the Curies grasped the central fact that radioactive behaviour was not chemical, that its energy and its ability to invade other substances were due to some property of the atoms themselves. These early researchers were quite unaware of the nature of the materials they were handling, and many of them, including Marie Curie and her daughter Irène Joliot-Curie, also a distinguished atomic physicist, died of their effects. The traditional view that atoms were the final, stable and indivisible constituents of matter – a view voiced by Newton and accepted by nineteenth-century scientists – was undermined by this discovery of the instability of radioactive elements, and of the ionisation of gases under electric charge.

In Cambridge between 1895 and 1897 Joseph Thomson conducted a series of experiments from which he concluded that emissions from the cathode (the nega-

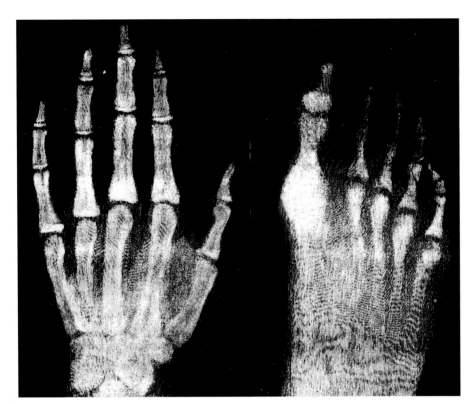

One of Röntgen's X-ray
photographs from 1896.
X-rays provided a vital
early clue to phenomenon
of radioactivity, and so in
turn to the structure of
the atom.

Marie Curie, who, with
her husband Pierre,
pioneered the study of
radioactive elements such
as radium and uranium.
Science Museum, London

tive electrode) in a vacuum tube were not rays at all, but behaved like particles – they were deflected by a magnetic field for example. They were not atoms, since he found these particles (soon to be called 'electrons') to be negatively charged and, most amazingly of all, he calculated that they must be less than one thou-

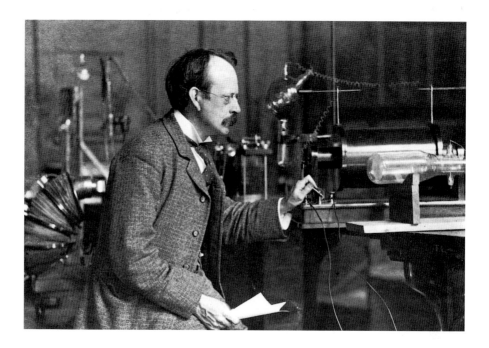

J.J.Thomson in 1896 with the electrical tube from which he deduced the existence of the electron.
Science Museum, London

sandth of the mass of the hydrogen atom, the lightest element known. The mass and charge of these particles appeared to be always the same, whichever gas was used in the tube, so that some fundamental particle common to all matter seemed to be discovered. In Thomson's excited words it was 'matter in a new state … this matter being the substance from which all the chemical elements are built up.' Thus the atom was not indivisible but it in turn consisted of still more minute, charged particles. But why should matter at this most elementary level be electrically charged? Was electricity truly diffused through space in the fields described by Maxwell, or was it concentrated into particles, and were charge and mass therefore fundamentally the same?

The Cavendish laboratory in Cambridge became for some years the centre of atom research, and the unsuspected instability of the atom, as revealed by Thomson and by the Curies, offered a pathway into its structure. One fundamental question was immediately apparent: if the electrons were all negatively charged, what counterbalanced them to give matter its electrical neutrality? Thomson at first conjectured that in the atom, positive and negative particles were all compressed together, in a famous image, like 'fruit in a pudding' although some were clearly capable of being chipped off, by electromagnetic force for example. In a series of historic experiments between 1898 and 1910, Ernest Rutherford (1871–1937) subjected Thomson's particles to a number of physical tests, as a result of which he succeeded in building a new and astonishing model of the atom, which he announced in 1911. The most decisive experiment was to beam radioactive particles through minutely thin gold foil; as expected almost all the particles penetrated, but a tiny proportion of them were reflected, prompting Rutherford's famous remark: 'It was almost as incredible as if you fired a 15-inch shell at a piece of tissue paper, and it came back to hit you.' Rutherford was driven to the conclusion that almost the entire mass of the atom consists of a dense nucleus, that this is surrounded by much smaller particles, distanced from the nucleus by a relatively huge empty space. His reflected particles had struck the

The classic model of the atom, initially developed by Rutherford and Bohr: the electrons orbit the nucleus as if in a miniature solar system, while the nucleus itself was later found to be composed of protons and neutrons.
Science Photo Library

nucleus, while the majority passed through the empty space. The first books which attempted to popularize atomic discoveries would describe the nucleus as the size of a pin-head in the centre of a vast cathedral, with the still tinier electrons orbiting in the dome. The outer particle, the electron is revolving so swiftly (at approximately one tenth of the speed of light) that the atom behaves like a solid, in the same way that the spaces between the blades of a high-speed turbine vanish for all practical purposes. The nucleus carries a positive electrical charge, the electron a negative charge, so that the whole atom is electrically neutral. This image of the atom as a solar system in miniature was over-simplified in many respects, but it has always remained a linchpin of modern physics. Research with atoms of different elements revealed that they have a different number of electrons and particles in their nuclei: suddenly the weights of the elements in the periodic table began to make sense: the characteristics of related elements derived from their atomic structure. To change the number of atomic particles would be to change the element, and this is exactly what Rutherford did when he disintegrated atoms of nitrogen and produced oxygen; the alchemist's dream of transmutation was finally achieved in a Cambridge laboratory in 1919. It was Rutherford who also realized that atomic processes at work within the earth were acting as a source of heat which would last for hundreds of millions of years, thus answering the nineteenth-century physicists' objection to Darwinianism. The structure of the atom as revealed by Rutherford, remains one of the central mysteries of nature: the principle, sanctified by science and by common sense, that matter is 'solid, massy, hard, impenetrable' was replaced by a vision of matter as little short of ghost-like, as necessarily composed largely of empty space because its nucleic material was so dense that a spoonful of it would weigh a million tons.

The great problem with Rutherford's model of the atom was this: according to classical physics, a charged particle moving in a curved path will emit electromagnetic radiation, causing a loss of energy. Thus the electron should spiral down into the nucleus, and the whole structure of the atom should be so unstable that it could not exist. This was the rock on which classical physics foundered: the physics of Newton and Laplace, of Kelvin and Maxwell, did not work at the sub-atomic level. The solution was proposed by the Danish physicist Niels Bohr (1885–1962), who applied to the atom a new kind of physics that had recently been developed in Germany – quantum theory. The founder of quantum theory was Max Planck (1858–1947), who devised the concept to explain major anomalies in the radiation of heat and light from certain heated bodies which were being studied. Attempts to chart a steady progression in the spectra and wavelengths of that radiation failed; it seemed, mysteriously, that energy was emitted or absorbed not on a continuous rising scale of values, as one would naturally expect, but on discrete levels. Planck divined that energy was exchanged in finite amounts, in packets which he termed quanta. Only by distributing these quanta of energy in statistical amounts over the whole radiation process, could the spectra for different temperatures be correctly predicted. Planck's constant is an inconceivably small number, which when multiplied by the radiation frequency, gives the energy of each of these discrete quanta. The formula is $E = h\nu$, where E is the energy, ν is the radiation frequency and h is Planck's constant. Planck's constant therefore gives a measure of energy multiplied by time, and is often referred to as a quantum of action. Planck's result was first announced in 1900, and in the following ten years physicists came to see it a powerful model for analysing not just energy, but charge or momentum. One of the first to embrace the concept was Einstein, who used it to explain the photoelectric effect by treating light as 'quantized' particles – later called photons. By 1916 Niels Bohr was postulating that the mysterious path of the electron is

Max Planck, the founder
of quantum theory.
Science Museum, London

quantized, that is, its orbit can lie only at discrete distances from the nucleus. Energy exchange at the atomic level results in the electron jumping from one orbit to another, absorbing energy as it moves farther from the nucleus, emitting energy as it jumps back towards the nucleus. The stability of the atom is explained because the electron cannot lose more energy than it has in the smallest orbit, where $n = 1$. Thus the solar-system model of classical mechanics, where electrons revolve in fixed paths like the planets, governed by Newtonian laws, has vanished, to be replaced by a series of possible orbits, to which the electron moves as it gives and receives energy. The speed and the scale on which these quantum actions occur would make it impossible to build a true, visualized model of the atom: it is describable only in mathematical patterns.

Even Bohr's model however was not flexible enough to deal with the heavier atoms having many electrons. In the 1920s quantum physics made dramatic progress by embracing the paradoxical principle that radiation, including the energy of the electron, behaves simultaneously like particles and waves. For example light, having been described as a wave by electromagnetic theory throughout the nineteenth century, was found to be analysable as discrete quanta of energy, now called photons. This duality was proposed intuitively by the French physicist Louis de Broglie, and given mathematical form by Erwin Schrödinger (1887–1961). Schrödinger's wave equations of 1926 treat electrons as waves which increase in energy and frequency level as they move farther from the nucleus. The energy levels of the electrons, and their transitions, were analysed the year before Schrödinger by Werner Heisenberg, using a quite different but mathematically equivalent model. The fundamental innovation contributed by Heisenberg, and accepted by most of the quantum pioneers, was the famous uncertainty principle, that the behaviour of atomic particles is so complex that their individual positions in space and time can never be precisely described, and they may only be predicted in terms of statistical probability. Heisenberg (1901–1976) insisted that the discontinuity of the quantum leap made it necessary to abandon completely the attempt to visualize the atom in conventional terms, and to think of it only mathematically. Mathematics and physics therefore do not claim to portray reality *as it is,* for its concern is only with what we can *say* about it. Added to this is the observation paradox: no test can be devised to determine the position of an electron, for any observation must entail some input of energy, which will cause its position to move. Like a coin spinning in the air, the question of heads or tails can only be decided by radically interfering with its movement. The observer interacts with his subject in a way which excludes the possibility of objective knowledge of matter at the deepest level. The extension of quantum mechanics to electrons moving at speeds approaching that of light, led Paul Dirac (1902–1984) in 1928 to postulate the existence of sub-atomic particles with charges opposite to those already known. The discovery in 1932 of the positron, which annihilates electrons and so acquired the reputation of being a form of 'anti-matter', was a triumph for Dirac's theory. The progress of particle research since then has revealed a complex world of sub-atomic particles resembling pure charges of high energy, which may be called into existence for a fleeting moment, but which have no separate being. The image – it cannot be called the structure – of the Rutherford-Bohr atom is probably as valid as the image of God in a Renaissance painting: it expresses a

Erwin Schrödinger, one of the seminal theorists of atomic structure in the 1920s.
Science Museum, London

human intuition rather than the reality itself. In place of reality we have now only probability, only the function of selected measurements.

The implications of quantum theory were not lost on its proponents. The assertion that the structure of matter was, at its deepest level, fluid, uncertain and beyond direct observation, was profoundly disturbing, and many were reminded of the ancient belief that nature was a Heraclitean fire, where change is the only reality. Men like Planck and Schrödinger were products of conservative, cultured families, well-versed in philosophy, and they were reluctant to follow the nihilistic direction in which their work seemed to be leading. The indeterminism latent in quantum theory was in conflict with their intuitive beliefs about nature. Planck was a man of unflinching integrity and moral force, remaining in Germany throughout the war and openly opposing Nazism; but the great international renown which protected him could not save his son from execution for his part in the July plot against Hitler. Schrödinger had a subtle mind that was drawn to oriental philosophies of wholeness, and he intuitively resisted the vision of indeterminacy inherent in quantum mechanics. Both Planck and Schrödinger retained their faith that human reason and its structures participated in some way in the laws of nature, and that the two should not therefore be at variance, as quantum theory seemed to show. Even Heisenberg himself, the apostle of uncertainty, came in his later life to be dissatisfied with particle physics, and to believe that nature cannot function without symmetries, and that these should in principle be discoverable. Nevertheless it must be emphasized that the validity of quantum mechanics has been demonstrated over and over again; the principle of discrete, quantized energy underlies our understanding of matters ranging from chemical bonding to cosmology, and it functions in all electronic technology. The most famous sceptic of quantum theory was no less a scientist than Albert Einstein (1879–1955), who inaugurated his own equally profound revolution in science, but who could not accept the indeterminacy of quantum mechanics.

Einstein was often pressed for a quick definition of his theories, and on one occasion he replied by saying that before relativity it was always assumed that if all matter vanished from the universe, space and time would remain, but that according to relativity this was no longer true; it would be hard to find a better non-technical summary of his achievement. Einstein's theories were published in several stages (quanta?) between 1905 and 1916, initially while he was without a position in academic physics, and this may be part of the reason why we have no precise idea how his thoughts turned into their highly original channels. The first phase of the theory of relativity appeared in a paper of 1905, in which Einstein drew attention to the problem of relative motion experienced within different frames of reference. Several of his postulates were counter-intuitive to our normal sense of time and space, but the essence of his argument was that events are defined by light, and that since light takes a finite time to travel, there can be no absolute agreement, between observers in different frames of reference, about the timing of events; there are in fact no simultaneous events. The most important consequence following from this is that space and time must be seen as a continuum, as space-time. In different frames of reference, strange discrepancies occur: at faster speeds, time moves more slowly, and mass increases. The culmination of this latter effect is that matter and energy are inter-convertible, that the mass of a body is a measure of its energy content, expressed in the famous formula $E = mc^2$; the energy released in this conversion is nuclear energy.

The effects of relativity are of course undetectable in our everyday experience, but they have important consequences at the atomic level, where the high speed of particles gives them a mass out of all proportion to their size. It is also revolutionary in the field of cosmology, where the inter-convertibility of matter and energy leads to the conclusion that light is subject to gravity, that it is curved around massive bodies such as stars. But if light defines events, indeed defines space, and if it is curved, it follows that the space-time continuum is not a series of Euclidean planes, but is curved; gravity can be seen therefore as the curving of space-time due to the presence of mass in the universe. If there were no mass-energy in the universe, there would be no space-time. The bending of light in the gravity-field of a star was verified in a famous experiment in 1919, organized by the British astronomer Sir Arthur Eddington. The result was that Einstein was enveloped in a tidal-wave of fame, and hailed as the author of a new vision of the universe. The very strangeness of his ideas added to his celebrity, and his mysterious paradoxes, for example that the universe must be considered finite but without boundaries, were endlessly discussed. It was commonly said that Newton's physics had been overthrown, but Newtonian physics still govern our terrestrial experience – they are all that is needed for aero-engineering and aero-navigation for example, or for space travel within the solar system. It is on the atomic and cosmological levels that Einstein's extraordinary ideas have proved so epoch-making. The last thirty years of Einstein's life were spent in the search for 'unified field theory' which would embrace gravity and electro-magnetism, the cosmos and the atom. He was profoundly distrustful of quantum theory, and felt that that it must be possible to build a model of physical reality as it is, and not merely a probability system. His criticism of Niels Bohr, that 'God does not play dice' has become famous; but perhaps Bohr's reply deserves equal weight, that 'we should not tell God how to organize the universe'.

COSMOLOGY

Of all the advances and discoveries of twentieth-century astronomy, those in the field of cosmology have made the deepest impression, both inside and outside the scientific community. Current ideas about the dimensions, the age and the probable origin of the universe have merged into a modern creation myth, overwhelming in its scale, and terrifying in its impersonal energy. Modern observational cosmology began with William Herschel's interest in nebulae (see above p.197) and his work was continued by his son John, who by 1864 was able to publish a catalogue locating some 5,000 of them. Even this was overshadowed by the work of the Danish astronomer Johann Dreyer, who by 1908 had published a massive catalogue of almost 15,000 nebulae. By this date too, first drawings and then photographs had revealed the curious disc-like or spiral shapes of many of the nebulae; yet their nature and significance was still quite uncertain. This situation changed with the inauguration of the new, giant telescopes in California, the 60-inch Mount Wilson reflector completed in 1908, and then the 100-inch of 1918. These instruments showed that some nebulae were indeed diffuse clouds of stellar dust, such as the famous Horsehead Nebula in Orion, but that the majority of them could be resolved into star fields. But how large were these star-fields, and why should stars form such well-defined disc-like groupings? The key to the problem was their distance, for some astronomers began to suspect that they were island-universes in their own right, but in the absence of measurable parallax how could their distance be gauged – whether they were within our own star-system or beyond it? Various means were tried to establish a ladder of distances, based on luminosity for example, but for many years the leading researchers in this field disagreed fundamentally about the scale of what they were observing. The

American Harlow Shapley (1885–1972)revised all previous estimates of the size of our galaxy, suggesting that it was around 100,000 light-years across, and Shapley believed that this was large enough to place all astronomical objects including the nebulae within the single system. His contemporary, Heber Curtis, disagreed, believing that our galaxy was smaller than Shapley claimed, and he was drawn, for both technical and intuitive reasons, to the view that the nebulae were island-universes far outside our own; he stated that the observable universe might extend to as much as 100 million light-years. During the years 1918–1924 observations with the large telescope continued, but the impasse prevailed.

It was another researcher who had arrived at Mount Wilson in 1919, Edwin Hubble (1889–1953), who devised a means of estimating distances to the nebulae, which convinced his colleagues and settled the controversy. In 1923 Hubble discovered in the Andromeda Nebula a type of star known as a Cepheid variable. These are stars which vary in brightness over a fixed period, and there is a constant ratio between their luminosity and their period; they function like natural, pre-set beacons. The distance to the relatively close Cepheids had been established by parallax measurement, and the period of the more distant ones enabled their absolute luminosity to be determined, from which their distance could be calculated. Hubble found no less than thirty-six such variable stars in the Andromeda Nebula, and he was able to determine that they must lie at almost one million light years distance. This was far beyond even Shapley's large-galaxy conception, and when Hubble's results were announced in December 1924, it was generally recognized that a landmark in cosmology had been passed, that a new scale and possibly a new structure for the universe had been revealed. The way forward lay in a closer study of the character of the galaxies, and of their distribution in space: how important were the different galactic shapes which had already been observed, and were the galaxies located at random, or was there a significant structure to the universe? Hubble played a leading part in both these fields, and over the following three decades he produced a classification system of galaxies, showing the many variations on the basic forms of eliptical and spiral, although he did not theorize about any possible process of evolution which would explain these forms. His other fundamental task was to build up a scale of distances, using natural markers such as variable stars wherever possible. By around 1929 Hubble had made observations which he believed extended the cosmic distance scale to some 250 million light years. Hubble was aware that his estimates carried a high degree of risk, and this was confirmed when the 200-inch reflecting telescope at Mount Palomar was commissioned in 1949. But it was found that, far from exaggerating the scale of the universe, all Hubble's extra-galactic distances had been *understated* by a factor of two.

The improving techniques of distance-estimation permitted a new approach to the problem first addressed by Herschel, namely mapping the universe in three dimensions. At first Hubble believed he had found a non-uniform pattern, for he identified a large region of the sky where no galaxies appeared, the so-called 'zone of avoidance' in the plane of the Milky Way. However he was later able to explain this is an aberration caused by a layer of diffuse interstellar dust which absorbed light, and if that was discounted, then the distribution of galaxies appeared to be uniform when taken on a sufficiently large scale. This observation was extended by the English astronomer E.A.Milne into a 'cosmological principle', that the universe is isotropic and uniform in every direction, a principle that was to have great importance in studying the dynamics of the universe. Later work on galactic mapping, since Hubble's death, has in fact revealed significant clusters or major groups of galaxies, which raises a serious question about this principle of

uniformity; whether these clusters emerged first or have arisen as existing galaxies drew together is still undecided.

By 1929 Hubble's study of the galaxies had yielded his most historic discovery: when their light was subjected to spectral analysis, it was found to be shifted to the red, a characteristic of a receding light-source. Moreover Hubble was able to show a constant ratio between distance and velocity of recession: the distant galaxies were receding from the earth, and the further they were from the earth, the faster was their speed; some of the galaxies observed by Hubble appeared to be moving at velocities up to one seventh of the speed of light. This discovery was widely likened to a new Copernican revolution, for in place of the eternal, unchanging, motionless heavens, there now appeared a universe of intense, explosive movement. Nor was there any reason to suppose that this recession was specifically from the earth, for the earth could not possibly be imagined to be the centre of the universe, rather this recession must be supposed to be observable anywhere, and be another feature of the cosmological principle. The conclusion seemed inescapable that all the galaxies in the universe were rushing away from each other at enormous speed, and that the universe was therefore expanding. The novelty of this vision was such that even Hubble himself had doubts about his findings, and thought that the recession might be merely apparent, and that

The galaxy M51, a magnificent spiral galaxy in the constellation Canes Venatici, photographed in 1950 as part of the Hubble research programme.
Carnegie Institute, Washington

[235]

perhaps some unexplained physical effect might be at work, so that the universe might truly be static after all. It is important to notice that everything in the universe is not expanding: the galaxies are not expanding internally, although they are in motion and they are evolving. The movement of Sirius which William Huggins found in 1868 was the motion of the star in relation to other stars, not part of the cosmic recession. It is the galaxies which are the large-scale units of the universe, and their distribution and behaviour is the key to cosmic structure and origins.

Although some astronomers shared Hubble's doubts, many theoretical physicists recognized that a vital new principle had been discovered, and they seized on Hubble's results as confirmation of some of Einstein's predictions. If the universe was expanding, the obvious questions were: What had it expanded from, and what was it expanding into? To take the second question first, the relativity of space to matter and light had now found a concrete expression: as the galaxies moved into ever-more remote space, so they defined that space and with it the structure of the universe. The paradox that the universe was at once finite but without boundaries, now made sense, as did the curvature of space, for the universe could now be imagined as an expanding sphere, in which the distance between *all* points was simultaneously increasing. Perhaps even more intriguing, if the film of the expanding universe was rewound, what happened at the beginning? Even before Hubble's results were announced, some physicists – the Russian Aleksandr Friedmann (1888–1925) and the Belgian Jesuit, Georges Lemaître (1894–1966) – had predicted, on the basis of Einstein's model, that the universe was non-static, and that the curvature of space was increasing with time. The only possible conclusion was that this expansion had a beginning in time, and in 1931 Lemaître introduced the quantum concept, that if we go back in time, fewer and fewer quanta of energy will be found, until all the energy of the universe is packed into a single quantum, a primeval atom of unimaginable density. This *Ur-atom* is surely a concept that would have delighted the German nature-philosophers of the eighteenth century.

Lemaître could go no further at that time, for no theory of nuclear processes was available to explain how the primeval atom might have evolved or exploded into today's universe. That element was provided by another group of theoretical physicists, notably the Russian, George Gamow (1904–1968), who by 1938 had concluded that the different elements in the universe could have become differentiated following a vast thermonuclear explosion, and he and his pupil, Ralph Alpher, later predicted that the radiation from this event should still be detectable. When Arno Penzias and Robert Wilson identified such radiation in 1965, it was seen as virtually proving Gamow's model. It is perhaps surprising that the most celebrated theory of twentieth-century science – the Big Bang origin of the universe – should not be associated with any single individual, as Newton is with gravity or Darwin with evolution, but Lemaître and Gamow were undoubtedly its two chief architects. The phrase Big Bang was first used as a term of derision in 1950 by the English astronomer, Fred Hoyle, who became the most famous critic of the theory, preferring instead a form of steady-state creation, in which the universe had no beginning, explosive or otherwise.

A logical consequence of the Big Bang theory is that it should be possible determine the age of the universe by calculating the known velocities and distances of the receding galaxies. In fact there are so many variables concerning the mass in the universe, and the constancy of the speeds that no unanimity has been achieved. The first post-Hubble estimate was two billion years, proposed by Eddington in 1932, but this has been revised steadily upwards, and there is no real reason to suppose that the current figure of fifteen billion years is final. The Big Bang is not a single theory, but a family of related ideas, with many competing accounts of how it might have occurred. Some years before the thermonuclear model of Gamow, an important theoretical advance had been made by the German astronomer Karl Schwarzschild (1873–1916), who studied the properties of dense stars. Schwarzschild imagined stellar matter increasing in density to the point where its gravity was so strong that nothing, neither light nor energy, could achieve the necessary velocity to escape from its surface. In mathematical language this was a 'singularity', a point-mass which is invisible but still detectable through its gravitational effects. This was the beginning of the theory of black holes, so important because they exemplify conditions where the normal laws of physics do not apply, but which may well have existed at the origin of the universe, and also perhaps at its end. Such a black hole is believed to exist at the nucleus of our Galaxy, which lies 6 degrees west of the star gamma Sagittarius – the arrow of Sagittarius points directly towards it. Black holes are one aspect of the enormous problem of dark or missing matter in the universe, a problem pioneered by the Swiss scientist, Fritz Zwicky, and where theoretical physics and observational astronomy interact and sometimes conflict. Zwicky (1898–1974) calculated the mass-luminosity ratio of the Sun and other single stars, and applied it over whole galaxies, with the astonishing result that galaxies appeared to possess fifty times more mass than their luminosity would require. This discrepancy was confirmed, and even increased, by other researchers, and it is now believed that invisible or diffuse matter permeates the cosmos on an enormous scale. This problematic dark matter bedevils all attempts to calculate the mass and structure of the universe, and the way it may have evolved; it is now believed for example that the galaxies are really spherical in form, with the solid matter settled into a stable disc-formation, which is surrounded by a vast corona of diffuse matter. Without a reliable estimate of the mass of the universe, it is proving impossible to resolve the problem of whether the universe is open or closed – whether its expansion will continue forever, or whether it will eventually slow, halt and go into reverse, falling back into a state of singularity. The visible matter in the cosmos suggests that its gravitational force is not sufficient to halt expansion, which thus can continue forever, yet theoretical results from Zwicky's onwards show the opposite, that the mass of the universe means that it must one day begin to contract and then implode on itself; at present there is an impasse, which neither observation nor theory can resolve.

The atomic revolution in physics had an enormous impact on astronomy by providing an explanation of the nature of stars and of solar energy. The fire in the Sun and stars which could burn for thousands of years had long been a mystery to astronomers – what was the source of their stability, why do they not simply burn up? The analysis of stars by spectral type had begun in the 1860s and by the early years of the twentieth century the famous Hertzsprung-Russell model had been devised, which classified stars by their luminosity and temperature. In this model an unmistakable line of progression appeared from intense heat and light, cooling into reduced activity and eventual extinction. It was understood that this model actually showed the different stages in the evolution of a star, but before the advent of atomic physics, no known energy-mechanism could account for this evolution, thought by then to extend over millions of years. Intriguingly, during the years 1850–1910 geologists' estimates of the age of the earth were much higher than physicists' ideas about the possible age of stars. But by 1917 Sir Arthur Eddington (1882–1944) had began to utilize Einstein's equivalence of matter and energy as the basis of a theory that star formation is the transformation of energy into matter, and that stellar energy is atomic energy. In his classic work *The Internal Constitution of the Stars,* 1926, Eddington argued that atomic fusion was occurring in the hot core of a star, releasing huge amounts of energy, while the cooler, outer layer of the star resists collapsing through gravity because it is bombarded by this intense energy from the core. Eddington gave an accurate quantitative basis to this theory, and showed that it was hydrogen fusing to helium which was at the heart of the process. The spectral composition of the Sun confirmed this, and its mass suggested a solar lifetime of ten billion years. Eddington's model was refined in 1938 when Hans Bethe (b. 1906)showed that a multi-stage cycle was operating in the Sun, whose beginning was the collision of carbon and hydrogen atoms, and whose end was helium and more carbon, so that the cycle could begin again. The classical chemistry of the nineteenth century was in a sense overturned, for the elements were not immutable after all; but the thermodynamic principle of the conservation of energy was demonstrated again, when once it was understood that matter and energy were inter-convertible, as Einstein had predicted. The structure and behaviour of the atom, so bizarre and unexpected, transpired to be the key to large-scale cosmic processes, one of whose incidental effects was to sustain life on this planet.

GENETICS

Two fundamental discoveries of the nineteenth century – the evolution of life's forms and the chemical basis of life's processes – have merged to place the science of genetics at the centre modern biology. It is often said that Mendel's work was rediscovered around the year 1900, but it is more accurate to say that his approach and his results were independently repeated by biologists such as Hugo de Vries (1848–1935) of Amsterdam, who then went on to formulate their own theories on the most difficult question of how the macro-process of evolution operated at the level of changes in the individual organism. De Vries agreed with Mendel that heredity worked through discrete factors, and he concluded that these were real, physical particles, which he named 'pangenes'. His experimental work turned him against a belief in natural selection as the true mechanism of evolution, and instead he proposed that mutations could occur spontaneously to produce sudden variations, which were often significant enough to amount to speciation. In this de Vries was in agreement with the controversial English Mendelian, William Bateson (1861–1926), whose views on morphological change verged on vitalism. The weakness in the 'mutationist' school was that they could offer no explanation of how spontaneous variations could possibly arise. It was Bateson who coined the word 'genetics' in 1906, but some of the basic termi-nology of the science was due to Wilhelm Johannsen (1857–1927) of Copenhagen, who in 1909 shortened de Vries's 'pangene' to simply 'gene', and who introduced the all-important distinction between the 'genotype', the genetic make-up with which an organism is endowed, and the 'phenotype', the mature organism as formed by environmental as well as hereditary factors. The progress from genotype to phenotype offers the enormous flexibility which accounts for the individuality of each creature.

The uncertainty about how genetic variations occur would continue for half a century, but by 1915 Thomas Hunt Morgan (1866–1945) had satisfied himself that the location of the genes was in the chromosomes in the cell nucleus. Mor-gan discovered the virtues of the fruit fly *drosophila* as a subject of genetic research: it reproduces rapidly and has only four large chromosomes. Morgan re-affirmed the belief that small genetic mutations arose constantly in the fruit-fly population, and that natural selective pressure filtered those which gave some adaptive advantage. Morgan thus began the rehabilitation of natural selection, which had seemed to be so threatened by the Mendelian discovery of the immutability of the genetic factors. It was realized that the inheritance pattern of a mature species such as man was more complex than the seven characteristics studied by Mendel, which represented the simplest possible case. In practice, although each factor may be subject to Mendel's laws, characteristics may actually be shaped by more than one gene, with the result that subtle variations constantly appear. One of Morgan's pupils, Theodosius Dobzhansky (1900–1975) continued the all-important synthesis of evolution and genetics by arguing that change within the chromosomes occurs first at a physiological level, and these are the mutations which may flourish or be lost by chance. But on a secular level 'the influence of selection, migration, and geographical isolation then mold the genetic structure of the populations into new shapes in conformity with the secular environment and the ecology, especially the breeding habits, of the species.' This approach was both statistical and ecological, and it was reinforced by the field-work of Ernst Mayr (b.1904) who proposed that geographical separation lay behind speciation, that the variations within in each gene pool would accumulate to the point where a new species emerged. In 1940 Mayr solved a long-standing conundrum by defining the species as a naturally interbreeding group, a definition which has been widely accepted. Morgan, Dobzhansky and Mayr were some of the leading figures of the 'new synthesis' in biology in which the process of evolution was shown to operate through micro-changes in individual genes, but it cannot be said that the functioning or the causes of these changes were yet understood. There was always a temptation to believe that evolution could work only if some feedback mechanism existed to connect the genes with the environment; if, for example, the climate became colder, a species might need to develop fur or fat to adapt, but how could the gene, a microscopic biochemical unit, analyse a climate change and take the necessary action? That the appeal of Lamarckism was not dead is shown by the strange story of genetics in the Soviet Union between 1930 and 1960, where Trofim Lysenko (1898–1976) led a movement which repudiated natural selection, and affirmed that crops and livestock could be selectively improved in one generation and that the improved characteristics would be directly inherited. It has often been suggested that this Soviet genetic doctrine was a reflection of the Marxist belief in the power of revolutionary change.

The understanding of how mutations or variations take place at the genetic level took a giant step forward with the unveiling of the molecular architecture of the gene itself. DNA now occupies an almost-mythical place in the popular mind, indeed it ranks with the Big Bang as one of the pillars on which our view of the universe and ourselves now rests, for they represent opposite ends of the scale of physical reality. Just as the Big Bang involves tracing the universe back through unimaginably vast tracts of time and space, so DNA involves reducing the fabric of life to inconceivably tiny, but apparently purposeful units. The analysis of the DNA (deoxyribonucleic acid) molecule by Francis Crick (b. 1916) and James Watson (b. 1928) was announced in 1953, and it had been preceded by research going back to the late 1920s, when the English bacteriologist, Fred Griffiths, found that genetic information could be transferred from one organism to another, and by 1944 Oswald Avery in the United States had identified DNA as the material involved. In 1951 Alfred Herschey and Martha Chase proved this by marking the DNA with radioactive tracers, and they showed at the same time that

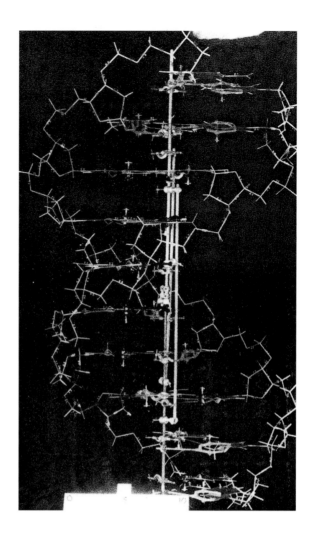

The original wire model
of the DNA molecule
built by Francis Crick and
James Watson in 1953.
The outer strings of sugar
and phosphates are linked
by bands of amines.
Athenium Press

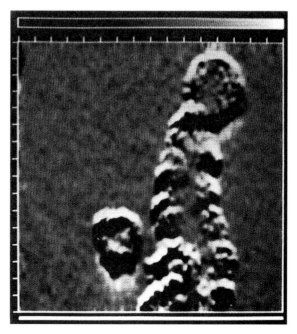

Photograph of DNA
made with a tunnelling
electron microscope; the
spiral structure is plainly
visible, while the strand
loops back upon itself.
Lawrence Berkeley Laboratory,
California

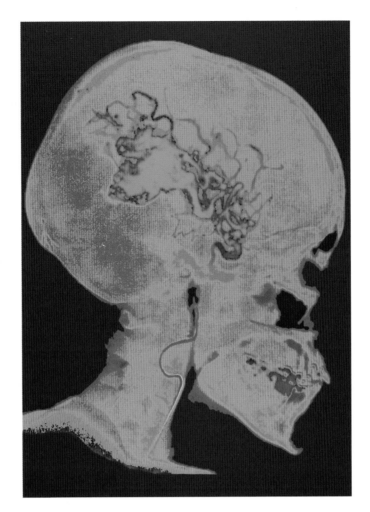

X-ray of a brain aneurism. The physical mapping of the human body has little more to achieve, but the physiology of consciousness, intellect and emotion still remains deeply mysterious.
Science Photo Library

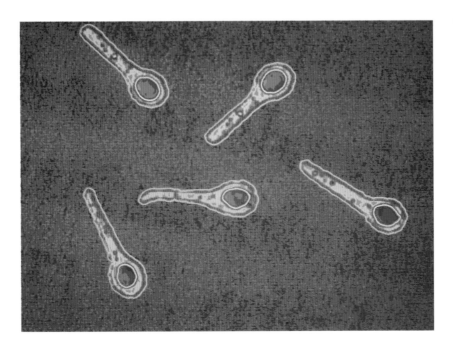

Tetanus bacteria: after centuries of ignorance about the source of disease, the hidden world of bacteria was first revealed by Pasteur: these organisms, and the even more minute viruses, hold the key to most human diseases.
Science Photo Library

Lightning over a city: the forces of nature harnessed by man, but never ultimately understood
Science Photo Library

DNA was replicating itself. DNA was known to consist of carbon, hydrogen, oxgyen, nitrogen and phosphorus, but its molecular structure was unusually complicated, and Crick and Watson owed their breakthrough to the X-ray crystallography carried out by Maurice Wilkins and Rosalind Franklin. Franklin in particular seems to have held all the necessary data, but hesitated as to the correct conclusion. The twisted ladder, or more scientifically the double helix, revealed by Crick and Watson showed how two strings of alternate sugars and phosphoric acids acted as a framework on which four relatively simple amines are arranged into almost infinitely varying sequences: this structure becomes a template, and its variety determines all the characteristics of all living organisms. The DNA itself is not the basic building material of life: it acts as a pattern or code which directs the construction of protein molecules in each cell, nor does it interface directly with protein, but sends the code via the messenger RNA, ribonucleic acid. This is achieved through the central process of molecular genetics – transcription from DNA to RNA, followed by translation from RNA to protein.

A gene is no longer regarded as an isolated entity, but rather as a portion of a chromosome, and it is more flexible than Mendel's 'factors'. Each human DNA molecule may contain millions of atoms, and each human cell contains around 100,000 genes; each gene is a specific sequence of amines, and in every gene the sequence is different. It has been estimated that the possible number of genetic sequences is 256×10^9, a vast number exceeding even those commonly used to express astronomical distances. The entire genetic sequence of every creature that has lived since life appeared on earth three billion years ago has not yet employed even half of the possible combinations. This means that evolution has further to go than it has yet gone. It also means that the human being is a wildly improbable creature, unless some process of design, intrinsic or extrinsic, is at work, for millions of variations have occurred in the genetic sequences and have either vanished or been perpetuated during the aeons of evolution of which man is the product. In the structure of the body as a whole there is huge paradox: all human cells contain all human DNA, which means that only a fraction of the DNA is used in any specific cell, as it differentiates into the many tissues and organs – bone, nerve, muscle, eye, brain, skin and so on. This requires a hugely intricate system of chemical activators and inhibitors to direct the growth of the cell in the appropriate way.

We know by observation that changes in genetic copying occur constantly, so that the building of proteins from amino acids does not always duplicate the DNA pattern with absolute precision, and the result is a mutation. We know that mutations can be caused by agencies external to the DNA, such as chemicals or forms of radiation including ultra-violet light or X-rays. But they can also arise spontaneously, as if they are copying errors. Most of these mutations will be negative or random, and they will not be conducive to adapting the organism in any positive way to its environment (consider the analogy of a sentence composed correctly of letters, words and punctuation marks: any random error caused by copying is highly unlikely to make the sentence clearer, and may indeed render it unintelligible). Some mutations may cause the organism to malfunction in a serious way; this is the 'inborn error of metabolism'. A major distinction has been found between the reproductive germ cells and the rest of the body cells, the soma cells. Mutations in soma cells will not be passed to offspring, but those in germ cells will be. August Weismann (above p.222) was intuitively correct in his germ-plasm theory. These mutations participate in natural selection by either disappearing, or, if they confer some adaptive advantage, becoming permanent. So the central, philosophical question about evolution is now reducible to this: is the mutation process merely random, and if so, is it conceivable that chance alone has guided the evolution of life forms from primitive amoeba to the complexity of the human brain? So central is the role of DNA in biological history, that the geneticist Richard Dawkins has spoken of life as inhering in the DNA, which uses the organism merely as a host to perpetuate itself. In this view, the entire process of evolution has a kind of secret history, which may be paraphrased as the expression of chemical codes; the forms of that expression are almost incidental, for it is the codes that are in control. This chemical cycle of replication, mutation and replication is probably even more more baffling in its impersonal drive than the idea of biological evolution used to be.

The revelation of the structure of DNA has been a landmark in our understanding of the mechanics of heredity and evolution, but such is the complexity of DNA that it has pushed further away our chances of unravelling the greatest biological problem of all – the origin of life. Ever since evolution came to dominate biological thought, it had naturally been assumed that scientific logic could eventually reverse the film, and by tracing the process back, in terms of morphology or chemistry, would discover how life began. The structure of proteins had yielded to chemical analysis since the time of Liebig, and it seemed feasible that amino acids might have formed naturally in the conditions that existed on earth several billion years ago. This concept lay behind the Miller-Urey experiments of the early 1950s, in which electric currents were passed through a supposedly primeval mixture of basic elements in water, with the result that several amino acids were indeed formed. Most of the excitement caused by these tests has now vanished with the realization that the gulf between a few amino acids and life is still so huge as to be unbridgeable. How proteins could have become organized into cells is impossible to say, and, above all, proteins cannot replicate themselves; DNA is as essential to life as amino acids. So there is now a double problem: to explain not only the emergence of complex protein molecules, but to explain the simultaneous appearance of DNA, or something very like it, without which any early proteins must simply have separated back into their constituent elements and vanished. The beginning of life is so enigmatic that the great Swedish chemist Svante Arrhenius (1859–1927) suggested that it could have had no origin on earth, but that spores or seeds had been carried through interstellar space by the pressure of light (an idea controversially revived in the 1980s by the cosmologist Fred Hoyle). Arrhenius's theory was a result of his view that the universe must be eternal, for the idea of a beginning must contradict the laws of thermodynamics, which are inviolable. Life could be seen as co-eternal with the universe, no more nor less of a mystery than its other material. This theory answers the problem of the origin of life on earth, but not the problem of the origin of life. Life has indeed been seen as a possibly unique exception to the second law of thermodynamics: that it represents an increase in order and complexity. This persuasive idea was advanced by Schrödinger in his book *What is Life?* Although powerfully argued, it is now considered mistaken, in that life is not a closed system, but is dependent on pre-existing thermodynamic cycles, and ultimately on the Sun: life increases in order at the expense of increasing disorder in the wider environment.

SCIENCE AND PHILOSOPHY REVISITED

The characteristic of twentieth-century science has been the dissolution, the fragmentation, the atomisation of reality. In every major discipline, it has been shown that reality is not what it seems. Matter is resolved into smaller and smaller particles; the universe recedes into structures and distances which defy the mind; while the definition of humanity seems to lie in impersonal, microscopic chemical patterns. The very earth beneath our feet, supposedly so solid, is now known to be in dynamic motion, as the continents drift and slowly reshape the map of the

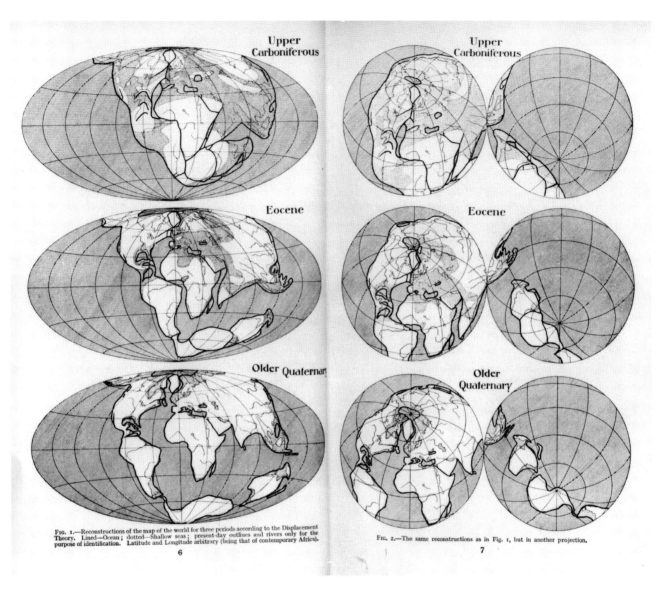

FIG. 1.—Reconstructions of the map of the world for three periods according to the Displacement Theory. Lined—Ocean ; dotted—Shallow seas ; present-day outlines and rivers only for the purpose of identification. Latitude and Longitude arbitrary (being that of contemporary Africa).

6

FIG. 2.—The same reconstructions as in Fig. 1, but in another projection.

7

world. Even deterministic systems have been shown to be unpredictable when their components are analysed in sufficient detail; this is the element of chaos, and it inheres in all systems, mechanical and natural, and especially in human society where probability reigns, hence the social sciences have a more problematic status than the other descriptive sciences. The revelation of hidden forces and hidden patterns has become the hallmark of modern science. The human psyche itself is also implicated in this process, for the unconscious has been so analysed and mapped that the mind is now seen to resemble an iceberg, nine-tenths hidden and invisible. The status of Freud's work as science is now in dispute, but its emergence during the destructive years between 1900 and 1940 serves as a paradigm

Continental drift proposed in the theory of Alfred Wegener, 1915. Wegener brilliantly deduced the existence of an astounding geological process; the later discovery of plate tectonics confirmed its mechanism.
The British Library
W.P.5708/4.

of the shifting foundations, the fragmenting reality, of twentieth-century thought. Behind modern science lie centuries of slow intellectual growth, in which it often seemed that philosophy was more important than knowledge. Science began to accelerate in the seventeenth century, when it detached itself to some extent from philosophy and took empirical knowledge as its field. Three centuries later, it is precisely because our knowledge has expanded so dramatically, that the questions of philosophy can no longer be avoided. The problem of design haunts modern science, but more bafflingly than it haunted the natural philosophy of the seventeenth or the eighteenth century, for behind the concept of design lies the problem of purpose. However deeply we probe the structure of matter, the chemistry of life, or the evolution of a galaxy, there is evidence of order, process, and the fulfilment of laws. But what end do these laws serve, and what law can explain existence itself? Science has disavowed the search for explanation in this ultimate sense, the reason why things are as they are and not otherwise. Perhaps there is no reason, perhaps this problem is an illusion of language, since matter can exist only in an ordered state, a state which we may call 'designed', and not in a state of chaos. But in that case we must be very clear about the status of scientific explanation – that it is ever-more-detailed description, description in which design is always implicit, but in which there is no designer, and no ulterior purpose to the design. All we can say is that the design works, and we cannot possibly test whether an alternative structure of matter, or an alternative form of human evolution, would work equally well. Science itself feels this shortcoming very acutely, and there is an urgency, almost a desperation, in contemporary claims that we are on the verge of a final breakthrough in our understanding the universe, the forces and the particles which compose it. Man cannot bear *not* to know, and he now knows so much that is secondary and contingent, that he feels that some primary truth must be within his grasp.

But why should the universe be comprehensible, and why comprehensible to man alone? The possibility that the human intellect somehow interacts with the laws of nature has recently come to seem very attractive, and has given rise to a new version of the macrocosm-microcosm theory of the middle ages. The so-called 'anthropic principle' suggests that the ultimate purpose of the universe requires that it should be understood; it suggests that man plays so a crucial role in this, that his evolution was essential to cosmic history, or indeed was even the purpose of cosmic history. The language is different, but the focus on man's unique role as the intermediary between earth and cosmos, was formerly the core of the religious impulse. For centuries, science flourished as a philosophy of nature which acknowledged that the final purposes of nature were hidden from man, without ever doubting that those purposes were real, and that they were under divine control. Science finally diverged from natural philosophy not when it proposed some new, secular purpose to nature, but when it turned aside from the search for purpose into ever-more-detailed description. The exact point at which this occurred is a matter of debate. In France it had occurred certainly by the late eighteenth century, when Laplace had, famously, no need of 'that hypothesis'. The formulation of the laws of thermodynamics or the theory of evolution might each be seen as the turning-point, for they showed the universe and the phenomenon of life both to be self-regulating processes. In the twentieth century, the revelation of the extraordinary nature of the atom has marked another stage in the enforced separation of science and philosophy, for the facts of atomic structure are so strange and paradoxical that no accepted theory of purpose could embrace them: they are unique and self-sufficient. But it is equally clear that the grim events of the twentieth-century have raised question marks over any facile belief in design or purpose in human history; in this sense, science has been only one among several forces which have dissolved the certainties of belief enjoyed by the nineteenth century. It is not scientific knowledge alone which lies behind Jacques Monod's claim that man 'knows at last that he is alone is the universe'. Virtually all religions of the ancient and modern world sprang from the conviction that the opposite is true, that a special relationship exists between man and the universe. Scientific knowledge and scientific techniques now dominate our civilization, but it is not yet clear which way they point in this matter, to a universe ruled by impersonal chance or one ruled by design, and if so, whose design it is. It would be the greatest irony in scientific history, if the answers emerged only with the rediscovery, within the heart of the atom or in the depths of the cosmos, of something which used to be called God.

Bibliography

This listing concentrates on classical science. In view of the current explosion in popular science publishing, I have not attempted to list the many excellent works which expound twentieth-century ideas such as quantum physics, genetics, and cosmology.

General Reference Works

C.C.GILLISPIE, ed. *Dictionary of Scientific Biography,* 16 vols. 1970-1980.
R.TATON, ed. *A General History of the Sciences,* 4 vols. 1963-1966.
C.SINGER et.al. *A History of Technology,* 5 vols. 1954-1958.
A.E.MCKENZIE, *The Major Achievements of Science,* 2 vols. 1973.
R.OLBY et.al., *A Companion to the History of Modern Science,* 1990.
H.SELIN ed. *Encyclopedia of the History of Science, Technology and Medicine in Non-Western Cultures,* 1997.

The Ancient and Medieval Periods

O.NEUGEBAUER, *The Exact Sciences in Antiquity,* 1957.
M.CLAGETT, *Greek Science in Antiquity,* 1976.
D.C.LINDBERG, *The Beginnings of Western Science,* 1992.
A.C.CROMBIE, *Augustine to Galileo: the History of Science AD400-1650,* 1952.
S.H.NASR, *Science and Civilization in Islam,* 1968.
S.H.NASR, *Islamic Science , an Illustrated Study,* 1976.
E.GRANT, *Planets, Stars and Orbs: the Medieval Cosmos,* 1994.
J.D.NORTH, *Chaucer's Universe,* 1988.
J.D.NORTH, *Stars, Mind and Fate: Essays in Ancient and Medieval Cosmology,* 1989.
J.E.MURDOCH, *Album of Science: Antiquity and the Middle Ages,* 1984.

The Renaissance and Scientific Revolution

A.C.CROMBIE, *Science, Art and Nature in Medieval and Modern Thought,* 1996
A.DEBUS, *Man and Nature in the Renaissance,* 1978.
D.P.WALKER, *Spiritual and Demonic Magic from Ficino to Campanella,* 1975
H.BUTTERFIELD, *The Origins of Modern Science,* 1957.
A.KOYRÉ, *From the Closed World to the Infinite Universe,* 1957.
I.B.COHEN, *Revolution in Science,* 1981.
D.GOODMAN & C.RUSSELL, *The Rise of Scientific Europe 1500-1800,* 1991.

A.R.HALL, *The Revolution in Science*, 1983.
S.DRAKE, *Galileo at Work: his Scientific Biography*, 1978.
A.TURNER, *Early Scientific Instruments, Europe 1400-1800*, 1987.
R.WESTFALL, *Never at Rest: a Biography of Isaac Newton*, 1980.
I.B.COHEN, *Album of Science: From Leonardo to Lavoisier*, 1980.

The History of Medicine

F.H.GARRISON, *Introduction to the History of Medicine*, 1913
and later editions.
C.SINGER & E.UNDERWOOD, *A Short History of Medicine*, 1962.
R.MARGOTTA, *An Illustrated History of Medicine*, 1968.
K.ROBERTS & J.TOMLINSON, *The Fabric of the Body: European Traditions of Anatomical Illustration*, 1992.
H.F.NORMAN, *One Hundred Famous Books in Medicine*, 1995.
R.PORTER, *The Greatest Benefit to Mankind: a Medical History of Humanity from Antiquity to the Present*, 1997.

Physics and Mathematics

C.BOYER & U.MERZBACH, *A History of Mathematics*, 1989.
H.EVES, *An Introduction to the History of Mathematics*, 1983.
C.J.SCHNEER, *The Search for Order: the Development of the Major Ideas in the Physical Sciences from the Earliest Times to the Present*, 1960.
I.B.COHEN, *The Birth of a New Physics*, 1985.
C.JUNGNICKEL & R.McCORMMACH, *Intellectual Mastery of Nature: Theoretical Physics from Ohm to Einstein*, 2 vols. 1986.
E.SEGRÉ, *From Falling Bodies to Radio Waves: Classical Physicists and their Discoveries*, 1984.
E.SEGRÉ, *From X-Rays to Quarks: Modern Physicists and their Discoveries*, 1980.
S.G.BRUSH, *The History of Modern Science: a Guide to the Second Scientific Revolution 1800-1950*, 1988.

Chemistry

A.J.IHDE, *The Development of Modern Chemistry*, 1984
W.H.BROCK, *The Fontana History of Chemistry*, 1994
C.J.SCHNEER, *Mind and Matter: Man's Changing Concept of the Material World*, 1988.
B.JAFFÉ, *Crucibles: the Story of Chemistry from Alchemy to Nuclear Fission*, 1968.
G.ROBERTS, *The Mirror of Alchemy*, 1996.
S.F.MASON, *Chemical Evolution: The Origin of the Elements, Molecules and Living Systems*, 1991.
M.CROSLAND, *Historical Studies in the Language of Chemistry*, 1962.

L.N.MAGNER, *A History of the Life Science*, 1994.

C.SINGER, *A History of Biology*, 1950

A.G.MORTON, *A History of Botanical Science*, 1981.

F.STAFLEU, *Linnaeus and the Linnaeans*, 1971.

W.LEY, *The Dawn of Zoology*, 1968.

P.BOWLER, *The Fontana History of the Environmental Sciences*, 1992.

P.BOWLER, *Evolution: the History of an Idea*, 1984.

F.D.ADAMS, *The Birth and Development of the Geological Sciences*, 1935.

A.HALLAM, *Great Geological Controversies*, 1989.

C.C.GILLISPIE, *Genesis and Geology*, 1959.

E.MAYR, *The Growth of Biological Thought: Diversity, Evolution and Inheritance*, 1982.

D.YOUNG, *The Discovery of Evolution*, 1992.

L.EISLEY, *Darwin's Century*, 1958.

J.NEEDHAM, *The Chemistry of Life: Lectures on the History of Biochemistry*, 1970.

M.FOURNIER, *The Fabric of Life: Microscopy in the Seventeenth Century*, 1996.

J.POSTGATE, *Microbes and Man*, 1969.

H.LECHEVALIER & M.SOLOTOROVSKY, *Three Centuries of Microbiology*, 1965.

L.PEARCE WILLIAMS, *Album of Science: The Nineteenth Century*, 1978

Astronomy

J.D.NORTH, *The Fontana History of Astronomy and Cosmology*, 1994.

C.WALKER ed. *Astronomy Before the Telescope*, 1997.

A.VAN HELDEN, *Measuring the Universe: Cosmic Dimensions from Aristarchus to Halley*, 1985.

J.D.NORTH, *The Measure of the Universe, a History of Modern Cosmology*, 1990.

T.KUHN, *The Copernican Revolution: Planetary Astronomy in the Development of Western Thought*, 1957.

C.WHITNEY, *The Discovery of Our Galaxy*, 1988

A.SANDAGE, *The Hubble Atlas of Galaxies*, 1963.

Index

Abu' Mashar, 52
Abu'l Qasim, 63
Academies, Scientific, 154
Adams, John Couch, 201
Adelard of Bath, 70
Afterlife, early beliefs in, 10, 17
Agassiz, Louis, 207
Age of the Earth, 142, 166, 174, 207
Agricola, Georgius, 104
al-Biruni, 63
al-Bitruji, 62
al-Din, Kamal, 61
al-Farghani, 55, 56
al-Ghazali, 58
al-Hakam II, 62
al-Haytham, 61
al-Idrisi, 62
al-Jazari, 60
al-Khayyami (Omar Khayyam) 60
al-Khazini 60
al-Khwarizmi, 55
al-Kindi, 55, 56
al-Mamun, 50, 51
al-Masudi, 62
al-Razi, 59
al-Sufi, 60
al-Tusi, 63
al-Uqlidisi, 55
Alberti, Leone, 101
Albertus Magnus, 75
Albinus, Bernhard, 173
Alchemy, 63, 83, 103
Alcuin, 67
Alexandria, 36
Alfonsine Tables, 79, 96
Algebra, 55, 147
Amici, Giovanni, 214
Anaesthetics, 220
Anaximander, 28-9
Anaximenes, 29
Angstrom, Anders, 196
Anthropic Principle, 244

Antiseptics, 220
Apollonius, 38, 43
'Arabic' numerals, 55, 71, 92
Aratus, 41, 46, 68
Archimedes, 37
Aristarchus, 42, 116
Aristotle, 29, 32ff, 50, 82
Arrhenius, Svante, 242
Asclepius, 35
Ashurbanipal, King, 21
Astrolabe, 57
Astrology, 20, 25, 26, 45, 52ff, 80, 87
Atmospheric pressure, discovery of,
   145
Atomic structure, 226ff
Atomic Theory, 186ff
Atomism, Greek, 29, 46, 83
Averroës (Ibn Rushd) 62, 82
Avery, Oswald, 239
Avicenna (Ibn Sina) 59, 82
Avogadro, Amedeo, 189

Bacon, Francis, 130ff, 182
Bacon, Roger, 74
Baer, Karl von, 214
Barberini, Maffeo (Pope Urban Vlll),
   127
Bateson, William, 238
Becquerel, Henri, 226
Bede, 67
Bell, Charles, 213
Belon, Pierre, 107
Beneden, Eduard van, 215
Bernard, Claude, 213
Berzelius, Jöns, 189
Bessel, Friedrich, 199
Bestiaries, 88
Bethe, Hans, 238
Big Bang, 236ff
Biochemistry, origins of, 220-1
Birunguccio, Vanocchio, 105
Black Death, the, 87

Black, Joseph, 179
Bock, Jerome, 106
Boerhave, Hermann, 171
Boethius, 49
Bohr, Niels, 230, 233
Bonnet, Charles, 166
Borelli, Giovanni, 170
Bouger, Pierre, 162
Boyle, Robert, 152-3
Bradwardine, Thomas, 84
Brahe, Tycho, 120ff
Broglie, Louis de, 231
Brown, Robert, 214
Brunfels, Otto, 106
Bruno, Giordano, 94, 119
Buckland, William, 205
Buffon, Comte de, 166
Bunsen, Robert, 199
Buridan, Jean, 79

Calendar, Egyptian, 19
Camerarius, Rudolf, 140
Campanella, Tomasso, 128
Campanus of Novara, 78
Cannizzaro, Stanislao, 190
Cardano, Girolamo, 91
Carnot, Sadi, 191
Carolingian Renaissance, 67
Casaubon, Isaac, 130
Cavendish, Henry, 185
Cell, discovery of, 214
Celsius, Anders, 179
Chastelet, Marquise de, 163
Chaucer, Geoffrey, 81
Cicero, 38, 41, 46
Circulation of the blood, 134ff
Clausius, Rudolf, 192
Clock, first mechanical, 85
Collins, John, 154
Comte, August, 212, 223
Condamine, Charles de la, 162
Continental drift, theory of, 243
Copernicus, 42, 90, 97, 113ff
Coulomb, Charles, 177
Crick, Francis, 239-241
Ctesibius, 38
Curie, Marie and Pierre, 226
Curtis, Heber, 234
Cusa, Nicholas of, 95
Cuvier, Georges, 174, 204

D'Holbach, Paul, 27, 180

Dalton, John, 186-188
Dante, 89
Darwin, Charles, 204ff
Darwin, Erasmus, 201
Davy, Humphrey, 188
Dawkins, Richard, 242
Decimal system, 179
Dee, John, 144
Delambre, Jean-Baptiste, 179
Democritus, 28-9
Derham, William, 180
Descartes, René, 132ff, 147, 162
Diderot and the *Encylopédie*, 180
Dinosaurs, 209
Dioscorides, 41
Dirac, Paul, 231
Divination in Babylonia, 22
DNA, 239ff
Dobzhansky, Theodosius, 239
Doppler, Christian, 201
Doyle, Arthur Conan, 220
Dreyer, Johann, 233
Dürer, Albrecht, 101

Eddington, Arthur, 233, 237
Ehret, Georg, 171
Ehrlich, Paul, 220
Einstein, Albert, 232-3, 236
Electromagnetic radiation, 195
Elements, chemical, 153, 185
Embalming, Egyptian, 17
Empedocles, 28-9
Enuma Elish (Babylonian epic) 23
Epicurus, 36, 39
Erasistratus, 40
Eratosthenes, 38
Ether, the, 162, 196
Euclid, 36, 39
Eudoxus, 33, 41
Euler, Leonhard, 163

Fabrici, Girolamo, 134
Fahrenheit, Daniel, 179
Fallopio, Gabrielle, 134
Faraday, Michael, 193
Faust legend, 118, 144
Fertilization, 34, 138, 214-5
Fibonacci, Leonardo, 71
Ficino, Marsilio, 93
Fischer, Emil, 221
Fizeau, Armand, 201
Flamsteed, John, 128

Fol, Hermann, 215
Fossils, early studies of, 141,174, 175, 204ff
Foucault, Jean, 199
Four Elements, 29, 33, 82
Franklin, Benjamin, 176
Franklin, Rosalind, 241
Fraunhofer, Joseph, 199
Frederick II, Emperor, 88
Fresnel, Augustin, 191
Freud, Sigmund, 243
Friedmann, Aleksandr, 236
Fuchs, Leonhard, 106

Galaxies, 234ff
Galen, 47, 109
Galilei, Galileo, 124ff
Galvani, Luigi, 172, 178
Gamow, George, 236
Gay-Lussac, Joseph-Louis, 188
Genetics, origins of, 221-2, 238ff
Geographical revolution, 91, 130, 159
Gerard of Cremona, 71
Gerard's Herbal, 108
Gerbert of Aurillac, 69
Germ theory of disease, 216ff
Gesner, Konrad, 107
Gilbert, William, 123, 144
Graf, Regnier de, 137
Gravity, Newtonian, 148, 151
Gray, Stephen, 175
Grew, Nehemiah, 140
Griffiths, Fred, 239
Grosseteste, Robert, 73
Guericke, Otto von, 145
Gutenberg, Johannes, 98
Guy de Chauliac, 87

Hale, George, 201
Hales, Stephen, 164
Haller, Albrecht von, 171
Halley, Edmond, 149, 152, 174
Hammurabi, code of, 22
Harrison, John, 179
Hartsoeker, Nicolas, 138
Harvey, William, 134ff
Hauksbee, Francis, 175
Heat, study of, 179, 191-2
Hegel, Georg Wilhelm, 223
Heisenberg, Werner, 231
Heliocentric theory, 42, 116
Helmholtz, Hermann, 192

Helmont, Jan Baptista von, 152
Heraclitus, 28, 29
Herbals, 88
Hermeticism, 93
Hero of Alexandria, 38
Herodotus, 18, 20
Herophilus, 40
Herschel, William, 195, 197-8, 233
Herschey-Chase experiments, 239
Hertwig, Wilhelm, 215
Hertz, Heinrich, 196
Hertzsprung-Russell model, 237
Hesiod, 28, 29
Hevelius, Johannes, 128
Hieroglyphics, Egyptian, 15
Hipparchus, 43
Hippocrates, 35
Homer, 28
Hooke, Robert, 136, 138
House of Wisdom, 51
Hoyle, Fred, 236, 242
Hubble, Edwin, 234
Huggins, William, 201
Humboldt, Alexander von, 205
Humours, bodily, 36, 48, 86, 101
Hunayn Ibn Ishaq, 51
Hunter, John, 172
Hutton, James, 174
Huygens, Christian, 128, 152, 155, 162
Hydrological cycle, 174
Hypatia, 48

Ibn Khaldun, 66
Ibn Tamiyya, 58
Ice Age, theory of, 207
Ingenhousz, Jan, 164

Jabir Ibn Hayyan (Geber), 65, 83
James of Venice, 71
Jenner, Edward, 172
Johannsen, Wilhelm, 238
Joule, James, 191

Kelvin, Lord, 192, 212
Kepler, Johannes, 122ff
Kirchhoff, Gustav, 199
Koch, Robert, 218, 220
Kühne, Wilhelm, 221

Lagrange, Joseph, 163
Lamarck, Jean-Baptiste de, 166, 211

Laplace, Pierre Simon, 164, 244
Lascaux cave paintings, 10
Lavoisier, Antoine, 185
Leeuwenhoek, Antoni van, 136
Leibniz, Gottfried, 134, 152, 162
Lemaitre, Georges, 236
Leonardo da Vinci, 98ff
Leverrier, Urbain, 201
Liebig, Justus von, 216
Linnaeus, Karl, 167-171
Lipperhey, Hans, 125
Lister, Joseph, 220
Locke, John, 153
Lockyer, Sir Norman, 12
Logarithms, 147
Loschmidt, Johann, 226
Lower, Richard, 135
Lucretius, 46
Lyell, Sir Charles, 207ff
Lysenko, Trofim, 239

Macrobius, 49
Mädler, Johann, 201
Magendie, François, 213
Magnetism, early studies of, 84, 144
Malpighi, Marcello, 136
Manilius, Marcus, 46
Mantell, Gideon, 209
Marci, Johannes, 145
Maricourt, Pierre de, 84
Martianus Capella, 49
Marx, Karl, 223
Maupertuis, Pierre de, 162
Maxwell, James Clerk, 191, 194-5
Mayer, Julius, 192
Mayow, John, 153
Mayr, Ernst, 239
Mean-speed theorem, 85
Méchain, Pierre, 179
Medina, Pedro de, 113
Megalithic astronomy, 12
Mendel, Gregor, 213, 221ff, 238
Mendeleyev, Dmitri, 190
Mersenne, Marin, 154
Merton Group, 84
Messier, Charles, 197
Mettrie, Julien La, 171
Michelson-Morley experiment, 197
Microscope, invention of, 136
Miller-Urey experiments, 242
Milne, E.A., 234
Mondino di Liucci, 87

Monod, Jacques, 245
Morgan, Thomas Hunt, 238
Mul Apin 25
Müller, Johannes, 213
Murchison, Roderick, 209
Music of the Spheres, 31, 124
Musschenbroek, Petrus van, 175

Napier, John, 147
Nature Philosophy, German, 181
Navigational science, 112-3
Nebulae, 197, 233
Needham, John, 166
Neptune, discovery of, 201
Neptunism, 174
Newcomen, Thomas, 182-4
Newgrange, 13
Newton, Sir Isaac, 148ff
Number systems , 'Arabic', 55
Number systems, Egyptian, 15
Number systems, Mesopotamian, 24
Nuñez, Pedro, 113

Observatories, Islamic, 63
Observatories, Modern, 233-4
Observatories, Scientific Revolution,
  122
Oersted, Hans, 193
Ohm, George, 193
Oken, Lorenz, 214
Olbers Paradox, 201
Omens, Babylonian, 22, 25
Optics, Greek, 39
Optics, Islamic, 61
Optics, Newtonian, 149
Optics, Medieval, 74
Oresme, Nicole, 79
Osiander, Andreas, 116
Oughtred, William, 147

Pacioli, Luca, 91
Paracelsus, 101, 119
Pascal, Blaise, 146
Pasteur, Louis, 216ff
Penzias and Wilson, 236
Periodic Table of Elements, 190
Perrin, Jean, 226
Peurbach, Georg, 96
Phlogiston, 171, 185
Photosynthesis, discovery of, 164
Physiology, 213
Piazzi, Giuseppe, 203

Planck, Max, 230
Plato, 31ff
Pliny, 46
'Pneuma', theory of, 32, 40, 48
Pollination, discovery of, 166
Porphyry, Tree of, 70
Porta, Giambattista della, 142
Posidonius, 45
Pouchet, Felix, 218
Preformation and epigenesis, 136, 138, 172, 214
Priestley, Joseph, 185
'Prime Mover', 34, 77
Printing, 98
Proust, Joseph-Louis, 188
Ptolemy, 39, 41ff
Pyramids, Egyptian, 16
Pythagoras, 28, 30

Quantum Physics, 230-232
Qibla, 52

Ramsden, Jesse, 179
Ramée, Perre de la, 122, 128
Raspail, François-Vincent, 214
Ray, John, 141
Redi, Francesco, 138, 166
Regiomontanus, 96
Rheticus, George, 116
Riccioli, Giovanni, 129
Richard of Wallingford, 85
Richman, Georg, 177
Roberts, Isaac, 201
Röntgen, Wilhelm, 226
Rømer, Ole, 128, 149, 179
Rutherford, Ernest, 228-230

Sacrobosco, 80, 112
Saint-Fond, Faujas de, 175
Saussure, Bénédict de, 207
Savery, Thomas, 182
Scheiner, Christoph, 127
Scheuchzer, Johann, 174
Schleiden, Jacob, 214
Schrödinger, Erwin, 231, 242
Schwann, Theodor, 214
Schwarzschild, Karl, 237
Sedgwick, Adam, 209
Semmelweiss, Ignaz, 220
Sevetus, Michael, 135
Sexuality of plants, 140
Shapley, Harlow, 234

Simpson, James, 220
Smith, William, 205
Snell, Willebrord, 147
Spallanzani, Lazzaro, 166, 217
Spectroscopy, 190, 199
Spencer, Herbert, 212
Spheres of the Heavens, 33, 39, 41ff, 78, 120
Spontaneous generation, 138, 166, 217-8
Sprengel, Conrad Christian, 164
Stahl, Georg, 171
Star-names, Arabic, 60
Steam power, 160, 182-4
Stellar astronomy, 197ff, 233ff
Stelluti, Francesco, 136
Stensen, Niels (Steno), 141
Stevin, Simon, 147
Stonehenge, 14
Stratigraphy, 205, 208
Swammerdam, Jan, 138-140
Swedenborg, Emanuel, 181
Swineshead, Richard, 84

Tartaglia, Niccolo, 91
Taxonomy, 141, 168ff
Telescope, invention of, 125
Telesio, Bernadino, 128
Thabit Ibn Qurra, 51
Thales, 28, 29
Thermodynamics, 191-2
Thierry of Chartres, 73
Thompson, Benjamin (Count Rumford), 191
Thomson, J.J., 226
Titius-Bode law, 203
Toricelli, Evangelista, 145
Tyndall, John, 222
Tyson, Edward, 141

Ulugh Beg, 62
Uranus, discovery of, 197
Ussher, James, 142

Vaccination, 172, 218
Vesalius, Andreas, 90, 108ff
Viète, François, 147
Virchow, Rudolf, 214
Vitalism, 171-2, 193
Vitruvius, 45, 46
Volta, Alessandro, 178
Voltaire, 162-3

Vries, Hugo de, 238
Vulcanism, 174

Watson, James, 239-241
Watt, James, 184
Wegener, Alfred, 243
Weissmann, August, 222, 241
Werner, Abraham, 174
West Kennet (prehistoric site) 14
Whewell, William, 222

Wilkins, Maurice, 241
William of Moerbeke, 73
Wolff, Caspar, 172
Writing, earliest, 15, 21

Young, Thomas, 191

Zij, 58, 79
Zodiac, 20
Zwicky, Fritz, 237